Earth from Above

USING COLOR-CODED SATELLITE IMAGES
TO EXAMINE THE GLOBAL ENVIRONMENT

Earth from Above

USING COLOR-CODED SATELLITE IMAGES
TO EXAMINE THE GLOBAL ENVIRONMENT

Claire L. Parkinson

LABORATORY FOR HYDROSPHERIC PROCESSES
NASA GODDARD SPACE FLIGHT CENTER

University Science Books
Sausalito, California

University Science Books
55D Gate Five Road
Sausalito, CA 94965

Fax: (415) 332-5393

Production manager: *Susanna Tadlock*
Copy editor: *Ann McGuire*
Designer: *Robert Ishi*
Compositor: *Wilsted & Taylor Publishing Services*
Printer: *Friesens Corporation*

This book is printed on acid-free paper.

Earth from Above constitutes a work of the United States Government
for which no copyright coverage shall exist in the United States.

Library of Congress Cataloging-in-Publication Data

Parkinson, Claire L.
    Earth from above : using color-coded satellite images to examine
the global environment / by Claire L. Parkinson.
        p.   cm.
    Includes bibliographical references and index.
    ISBN 0-935702-41-5
    1. Earth sciences—Remote sensing. 2. Artificial satellites
    in earth sciences. I. Title.
    QE33.2.R4P378   1997
    550–dc21   97-7657
                                            CIP

Printed in Canada
10 9 8 7 6 5 4 3 2 1

Dedicated to the memory of my father,
C.V. Parkinson,
whom I miss very much.

Contents

Preface

The advent of satellite technology in the second half of the twentieth century has provided mankind with an amazing new ability to see the Earth-atmosphere system in a way never before possible. With satellites, the most remote regions of the oceanic and continental surfaces can be viewed as readily as the least remote. Data can be obtained from the entire Earth surface and, from some satellites, this can be done every day or at least every few days. Additionally, atmospheric data can be obtained from all levels of the atmosphere, over any geographic location, and with updates possible on a frequent basis. Satellite data can thus reveal the state of the Earth-atmosphere system and many of the changes occurring within it.

Of course, for the satellite data to reveal anything, the user or reader of those data must be able to understand them. For many of the satellite data, scientists and computers greatly aid this process by converting the streams of raw data into color-coded maps of geophysical quantities of interest. The central objectives of this book all relate to increasing the understanding of satellite data and the maps or images created from them. These objectives are: to familiarize the uninitiated with satellite data and with the reading of color-coded satellite images of the Earth; to instill a sense of how the raw data are converted into information about the Earth and the atmosphere; to instill a sense of the range of information being collected about the Earth through satellites and a sense of the imperfections and cautions of which the users of satellite data should be aware; and to instruct on six of the important variables in Earth-atmosphere studies that are now being examined with data from satellites—atmospheric ozone, polar sea ice, continental snow cover, sea surface temperature, land vegetation, and volcanoes. These six variables come nowhere near to covering the full range of important topics being examined with satellite data, but they illustrate the types of information that such data are revealing about the Earth-atmosphere system. Furthermore, for each variable, one type of satellite data is highlighted, and this choice also is illustrative only, as each variable can be examined from several different satellite instruments. However, by learning to read the illustrative examples, an individual should become capable of reading all well-labeled, color-coded satellite images, not just the specific examples.

The book includes nine chapters—an introduction on visible data; a chapter on radiation, which is what each of the satellite instruments actually measures; chapters on each of the six selected Earth-atmosphere variables; and a concluding chapter that ties together various points made earlier and broadens the discussion to the strengths and limitations of satellite Earth observations in general, with mention also of improvements to be expected during the next decade. Each of the six Earth-atmosphere topical chapters is divided into three sections: an introduction explaining the highlighted variable and giving some of the reasons for its importance, a description of how information about the variable is obtained from the satellite instrument chosen for the illustrative results, and a section presenting the satellite images and describing aspects of what they show. Each section in each chapter concludes with a list of questions to encourage the reader to review the text and, for sections with satellite images, to examine the images and find answers from them. A list of additional readings near the end of the book can assist readers interested in finding further information about the six Earth-atmosphere variables, the satellite data illustrated in the text, and other relevant satellite data products.

By the time a person has completed the book, he or she should be familiar with several important topics in Earth-atmosphere studies and quite capable of reading color-coded satellite images, not just regarding the six illustrated variables but also regarding any other variable, and not just from the illustrated data sets but from a far wider range of satellite data. It is hoped that readers will find the book to be informative and fun and that, when they are done, they will be undaunted by any well-labeled satellite images they might encounter.

Acknowledgments

This book has benefited from the help of many people, both in the generation of the figures and through discussions and suggestions on the written text.

First and overwhelmingly foremost, I thank Jamila Saleh of Hughes STX for her help in putting into final form all of the color-coded satellite images, all of the location maps, and several of the schematic diagrams. Jamila worked closely with me on each of the products, in a process that often required many iterations and considerable patience and expertise on Jamila's part. She was a joy to work with.

Others who assisted with the satellite imagery, all also very much appreciated, are:

- Frank Corprew of the Goddard Distributed Active Archive Center (DAAC), who provided a CD-ROM containing ozone data.

- Liz Johnson and Rosanna Sumagaysay of the Jet Propulsion Laboratory (JPL) at the California Institute of Technology, who provided sea surface temperature data from the NASA Physical Oceanography DAAC via the world wide web.

- Jim McManus of Hughes STX, who helped in the reading and reformatting of vegetation data from a CD-ROM data set.

- Rich McPeters of Goddard Space Flight Center and Eric Beach and Ian Sprod of Hughes STX, who identified world wide web sites containing volcano data and described some of the data processing. Eric further helped through discussions regarding ozone calculations.

Others who assisted with the schematic diagrams and photographs, again very much appreciated, are:

- Mary Cleave and Jim Garvin of Goddard Space Flight Center, Charlotte Griner of Hughes STX, Lonnie Thompson of Ohio

State University, and Koni Steffen of the University of Colorado, each of whom provided requested photographs. Charlotte provided Figure 1.1 from the NASA archives; the others provided personal photographs (Figures 1.2, 4.1e, 5.2, and 8.4), identified in the figure captions.

- Mark DeBord and Debbie McCallum of the Goddard Photographics Services Team, who took great care in creating high-quality prints from a variety of photographic slides.

- Mary Pat Hrybyk-Keith and Debbi McLean of the Goddard Computer Graphics Facility, who converted my sketches of numerous schematic diagrams into professional products; Winnie Humberson, who similarly helped with Figure 2.3; and Freya Tanz, who similarly helped in creating drawings used in Figure 4.2.

- Kerstin Demko, who posed for Figure 5.1d, many years ago.

Of course, of primary importance in a scientific product is the accuracy of the science. In view of the range of topics covered, it was particularly important to seek out experts in the individual subject areas, and I thank many individuals for sharing their expertise both through written reviews and through discussions. The following experts kindly took the time to go through carefully the chapter in his or her area of expertise and to offer many valuable comments and suggestions:

- Walid Saleh of the Applied Physics Laboratory at Johns Hopkins University, who reviewed the radiation chapter (Chapter 2) and provided source materials,

- Susan Solomon of the National Oceanic and Atmospheric Administration, who reviewed the ozone chapter (Chapter 3),

- John Walsh of the University of Illinois, who reviewed the sea ice chapter (Chapter 4),

- Dorothy Hall of Goddard Space Flight Center, who reviewed the snow chapter (Chapter 5),

- Bill Lau of Goddard Space Flight Center, who reviewed the sea surface temperature chapter (Chapter 6),

- Fred Huemmrich of Hughes STX, who reviewed the vegetation chapter (Chapter 7) and provided source materials and information on the vegetation data set,

- Peter Mouginis-Mark of the University of Hawaii, who reviewed the volcanoes chapter (Chapter 8), and

- Ian Sprod of Hughes STX, who also reviewed the volcanoes chapter (Chapter 8).

Additionally, George Fisher of Johns Hopkins University kindly and patiently went through the entire manuscript and offered many valuable suggestions from the perspective of a scientist and professor planning to use the book in a university setting. Linda Davis of Wilde Lake High School in Clarksville, Maryland, similarly reviewed the first five chapters, although from the perspective of a teacher planning to use portions of the book in a high school setting. Jamila Saleh of Hughes STX went through Chapters 2, 3, and 4, checking the accuracy of statements regarding the computer processing; and Dorothy Hall of Goddard Space Flight Center went through the Preface and Chapters 1 and 2, from the perspective of a remote-sensing scientist, in addition to Chapter 5 in her specific area of expertise. All these efforts are much appreciated.

I also thank Sirpa Hakkinen of Goddard Space Flight Center for discussions regarding deep-water formation and the impact of sea ice, Inez Fung of the University of Victoria for correspondence regarding global carbon budgets, Richie Williams of the United States Geological Survey for discussions regarding volcano impacts and the loan of volcano slides, and Lori Glaze of the Universities Space Research Association and Goddard Space Flight Center for discussions regarding volcanoes.

Finally, the following people also helped in various ways and are greatly appreciated:

- Milt Halem of Goddard Space Flight Center, for encouragement and general scientific discussions,
- Jean Harris of the Pike's Peak Library District, for encouragement and discussions on several of the figures,
- Jerry Soffen of Goddard Space Flight Center, for taking the photograph of the author that appears on p. 176,
- Winnie Humberson of Hughes STX, for draft layouts of two of the chapters,
- Bob Ishi, for the final design of the book,
- Susanna Tadlock of Susanna Tadlock and Associates, and Bruce Armbruster and Jane Ellis of University Science Books, for their efforts at many stages during the production process,
- The National Aeronautics and Space Administration (NASA) and NASA's Goddard Space Flight Center, for employing me and providing the resources to do the work,
- The Earth Observing System (EOS) Project, and especially Michael King of Goddard Space Flight Center, for support and funding to do the work, and
- Bruce Armbruster of University Science Books, for publishing the book.

Earth from Above

USING COLOR-CODED SATELLITE IMAGES
TO EXAMINE THE GLOBAL ENVIRONMENT

Figure 1.1.
The Earth as photographed by the crew of the Apollo 17 spacecraft on their
way to the Moon, December 1972. Africa dominates the top left quadrant of the
photograph; the island of Madagascar lies just above and to the right of the
center of the photograph; and the white, ice-and-snow-covered continent of
Antarctica is partly visible, although somewhat obscured by cloud cover, at the
bottom of the photograph.

CHAPTER 1

Introduction:
Visible Images from Space

Many images from space can be readily understood by almost everyone, irrespective of scientific training. Among these are the photographs that the Apollo astronauts took on their journeys to and from the Moon, showing the sphere of the Earth isolated against the dark background of space (e.g., Figure 1.1), and the photographs that the Space Shuttle astronauts take from much closer to the Earth, during their missions in Earth orbit (e.g., Figure 1.2). These pictures show the dark oceans, the white clouds, and easily identifiable continental boundaries in areas unobscured by clouds.

Satellite images shown during television weather reports are also usually easily understood. These images come from the data of satellites that orbit much closer to the Earth than the Apollo astronauts were when they took photographs such as the one in Figure 1.1, and, partly as a result, they generally show a much smaller portion of the Earth, often centered on the local region of interest to the television audience. Figure 1.3 shows a satellite image similar to those shown on many weather reports. As in Figures 1.1 and 1.2, clouds and coastal boundaries are easy to recognize, although the ease of recognizing coastal boundaries has been greatly enhanced in Figure 1.3 by having the boundaries artificially overlain in yellow on the original satellite image.

Some other images from space show smaller regions and much greater detail. For instance, Figure 1.4 shows a satellite image of ice floating in the ocean near the continent of Antarctica. Individual floes of ice can be seen, and hence their sizes and shapes can be determined and analyzed. When the same floes are recognized in images taken at different times, the dis-

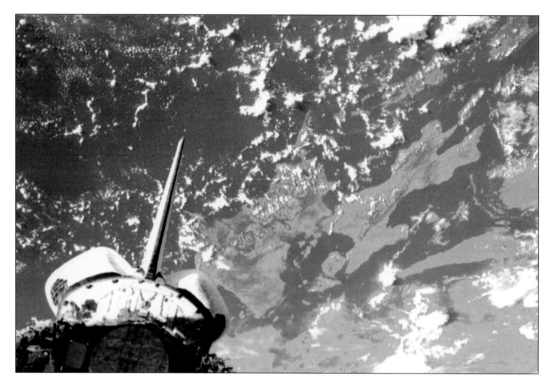

Figure 1.2.
Photograph taken from the Space Shuttle Atlantis while 330 kilometers over the Philippines, May 1989. The aft end of the Shuttle is visible in the lower left; and also visible are the Philippine islands of Bohol at the center of the far right, Cebu below Bohol, Leyte to the left of Bohol, and Samar to the left of Leyte and the upper right of the Shuttle. The image is oriented with north to the lower left. [Photograph kindly provided courtesy of the photographer, astronaut Mary Cleave.]

tances the floes have moved can be calculated, enabling information to be obtained about ice motions and sometimes also about the ocean currents and the winds that are causing the ice to move.

A primary reason that all the pictures in Figures 1.1–1.4 are so easy to understand is that they were taken with *visible* light, or, equivalently, visible radiation. Visible light is the type of radiation that human eyes can see (hence the adjective *visible*). Satellite instruments, similarly to standard, hand-held cameras, can be constructed to transmit information from visible light, resulting in images such as those in Figures 1.3 and 1.4. Satellite instruments, however, can also be constructed to transmit information from other types of radiation, often resulting in images that are not nearly as readily understood as Figures 1.3 and 1.4 but that provide considerable additional information about the Earth system.

Figure 1.3.
Hurricane Luis as its center passed just northeast of Puerto Rico, heading northwest, September 6, 1995. Luis, dominating the lower right of the image, contains a prominent "eye" and the distinctive counterclockwise spiral pattern of Northern Hemisphere hurricanes, as air swirls inward toward the low pressure in the center of the storm. Florida is in the center left portion of the image, and Cuba is the largest land feature in the bottom left quadrant. Luis caused an estimated $2.5 billion in property damage in the eastern Caribbean and killed 17 people. The image was obtained from the Defense Meteorological Satellite Program (DMSP) Operational Linescan System. GMT: Greenwich Mean Time. [Image used courtesy of the National Geophysical Data Center, Boulder, Colorado.]

To illustrate why it might be useful to examine information from types of radiation other than visible radiation, Figure 1.5 presents a Landsat visible image of sea ice under cloudy conditions. Notice that the clouds between the surface and the satellite severely reduce the amount of information the satellite data reveal about the ice conditions at the surface. Such images can be quite appropriate for studies of clouds, but in the presence of a heavy cloud cover they tend to be very inappropriate for studies of sea ice or other surface features.

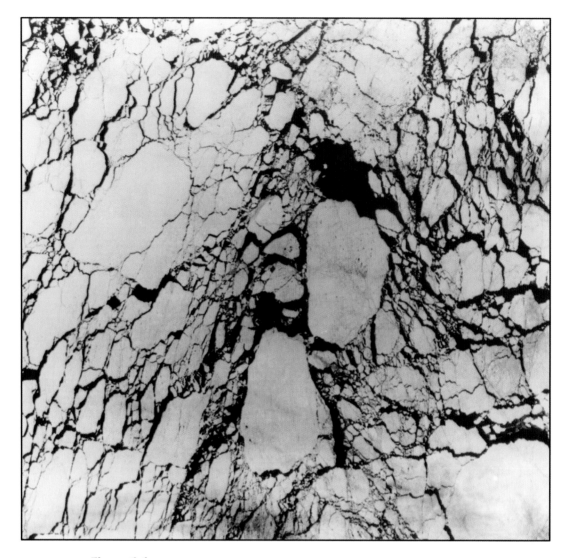

Figure 1.4.
Landsat image of sea ice in the Weddell Sea off the coast of Antarctica, November 17, 1973. The imaged area is approximately 185 kilometers on each side.

Fortunately, there are some types of radiation for which clouds do not obscure the underlying surface. Consequently, satellite instruments constructed to measure those types of radiation can record conditions at the surface even with an intervening cloud cover. This is a major reason for turning to nonvisible radiation for the examination of features at the Earth's surface that vary over time.

Another reason for sometimes preferring data from nonvisible radiation is that a satellite instrument measuring only visible radiation cannot obtain data for most variables during the darkness of night, in view of the

Figure 1.5.
Landsat image of clouds and sea ice in the Arctic, September 5, 1972. Sea ice is readily visible mainly in the left fifth of the image, the sea ice in the remainder of the image being largely obscured by clouds. The imaged area is approximately 185 kilometers on each side.

Sun's being the dominating source of visible radiation at the Earth's surface and in the Earth's atmosphere. With some of the nonvisible radiation types, measurements can be made irrespective of whether it is day or night, light or dark, summer or winter.

Although the discussion thus far has highlighted advantages and disadvantages of visible radiation, each of the other types of radiation also has its own sets of properties and its own advantages and disadvantages for use in satellite observing systems. The amount of information that can be gained about the Earth and its atmosphere through satellite observations depends very much on how well these sets of properties become understood and used; for it is only radiation, and not actual samples of the Earth or the atmosphere, that a satellite receives and transmits.

Because of the importance of radiation to the vast amount of information now being gathered about the Earth through satellite data, a knowledge of some of the basic facts about radiation is critical for an informed understanding of satellite observations. For this reason, the next chapter discusses some of these basics, prior to the presentation in subsequent chapters of illustrative, satellite-derived information on specific key variables in the Earth–atmosphere system.

Review Questions

1. **a.** On Figure 1.1, which continent's coastline is most clearly delineated, Africa's or Antarctica's?
   b. Why?

2. Of Figures 1.1, 1.2, 1.3, 1.4, and 1.5, which one was taken at the greatest distance from the Earth?

3. Of Figures 1.1, 1.3, and 1.4, which one shows the finest-scale details at the Earth's surface? (*Note:* The Earth's surface refers to ocean and ice surfaces as well as land surfaces.)

4. What is the name of the type of radiation that normal human eyes are able to see?

5. Name an advantage of using visible radiation to record images of the Earth.

6. Name two conditions under which visible radiation would not be the best type of radiation for satellite detection of conditions at the Earth's surface.

7. Which of the following four sets of conditions is preferable when using satellite visible imagery to study phenomena at the Earth's surface: cloudy daytime conditions, clear daytime conditions, cloudy nighttime conditions, clear nighttime conditions?

8. Which of the four sets of conditions in Question 7 is preferable when using satellite visible imagery to study clouds?

CHAPTER 2

Radiation

Introduction

All objects, including people, desks, books, the solid ground, the liquid oceans, and the gaseous atmosphere, give off radiation. Major differences exist in the types and amounts of radiation given off, but none of the differences alters the fundamental fact that all objects give off radiation. Furthermore, all the radiation given off travels through space at the same speed, approximately 300,000 kilometers per second. This is about 186,000 miles per second, or 11 million miles per minute, a speed unimaginable to most people. Indeed, as determined by Albert Einstein in 1905, no object can travel faster than the 300,000-kilometer-per-second speed of radiation through space. Phrased more familiarly: no object can travel faster than the speed of light. (Visible light, as mentioned in Chapter 1, is one of many types of radiation.) The speed of radiation is less when the radiation travels through media other than the emptiness of space; but the speed through air, which is the medium of most relevance to this workbook, is close to 300,000 kilometers per second.

All the various types of radiation can be distinguished by a property termed the *wavelength* of the radiation. Radiation travels through space in a manner similar to the motion of a wave, and the wavelength of any particular type of radiation is the distance from one peak of the wave to the next peak of the wave. Schematically, the wavelength is the distance labeled λ in Figure 2.1.

Visible radiation encompasses a span of wavelengths ranging from about 0.38 micrometers (which is 0.00000038 meters, or approximately 0.000015 inches) for light with a violet color to about 0.78 micrometers (0.00000078 meters, or approximately 0.000031 inches) for light with a red

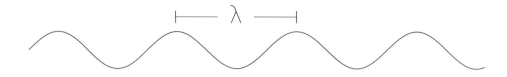

Figure 2.1.
A schematic wave of radiation, with the wavelength λ indicated.

color. The other colors of the rainbow have intermediate wavelengths between those for violet and red light, with the colors of the visible spectrum merging from violet to blue to green to yellow to orange and finally to red (Figure 2.2).

The Sun gives off (or *radiates*) a tremendous amount of visible radiation, whereas the Earth radiates very little radiation in visible wavelengths. It is the visible radiation from the Sun that dominates in allowing us to see objects outside during daylight. In contrast, at night humans have difficulty seeing anything terrestrial, except under the light of a bright Moon (which is actually visible light from the Sun, reflected off the Moon) or with the help of artificial light, such as from a light bulb or a flashlight. Both the light bulb and the flashlight are inventions that give off visible light through the use of electricity. Stars other than the Sun also give off visible light, and some of this light can be seen from the Earth.

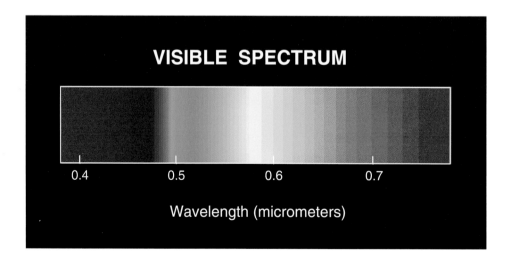

Figure 2.2.
The wavelengths and approximate corresponding colors of the visible spectrum.

These stars are so distant, however, that the amount of light reaching the Earth from them is very small compared with the amount coming from our closest star, the Sun. Therefore, they are not visible in daylight and do not provide much light for us at night.

Although the Earth and its atmosphere do not give off much radiation at visible or shorter wavelengths, they do radiate considerable amounts at other, longer wavelengths, a point that will be expanded upon and illustrated in a more general context later in the chapter. The radiation given off by the Earth–atmosphere system and the Sun's radiation reflected by the Earth–atmosphere system are the two central phenomena that allow the Earth to be studied from data collected by satellite instruments.

Review Questions

1. Name a familiar object that does not give off radiation.

2. How fast does radiation travel though space?

3. What is the limiting speed that no object can exceed, according to the theory of Albert Einstein?

4. What is the meaning of the "wavelength" of a wave of radiation?

5. **a.** Which color has the shortest wavelengths for visible radiation?
   b. Which color has the longest wavelengths for visible radiation?

6. What body radiates most of the visible radiation seen by humans?

7. Does most of the radiation given off by the Earth–atmosphere system have wavelengths that are longer than those of visible radiation or wavelengths that are shorter than those of visible radiation?

The Electromagnetic Spectrum

When the different types of radiation are arranged according to wavelength, the full sweep of radiation is termed the *electromagnetic spectrum* (Figure 2.3). In addition to visible radiation, some of the other familiar types of radiation are X-rays, microwaves, and radio-band radiation, with such familiar applications as the use of X-rays in medical examinations, the use of microwave radiation for heating food in microwave ovens, and the use of radio-band radiation for transmitting radio and television signals. As shown in the chart of the electromagnetic spectrum in Figure 2.3, X-rays have wavelengths shorter than those of visible light, whereas microwaves and radio-band radiation have much longer wavelengths. Immediately above visible radiation in Figure 2.3, with wavelengths shorter

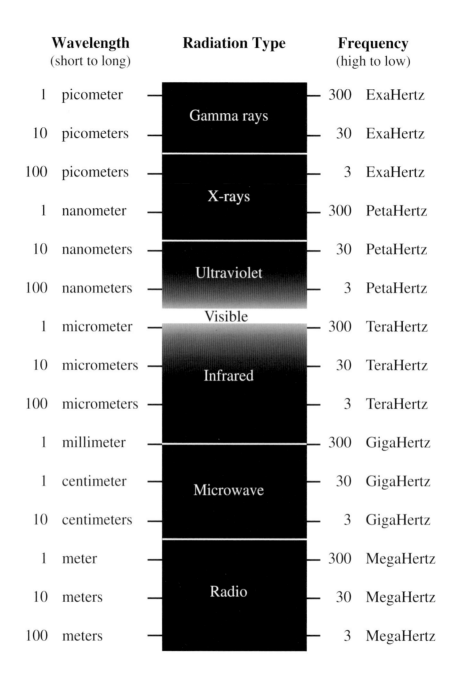

Wavelength (short to long)	Radiation Type	Frequency (high to low)
1 picometer	Gamma rays	300 ExaHertz
10 picometers		30 ExaHertz
100 picometers	X-rays	3 ExaHertz
1 nanometer		300 PetaHertz
10 nanometers	Ultraviolet	30 PetaHertz
100 nanometers		3 PetaHertz
1 micrometer	Visible	300 TeraHertz
10 micrometers	Infrared	30 TeraHertz
100 micrometers		3 TeraHertz
1 millimeter		300 GigaHertz
1 centimeter	Microwave	30 GigaHertz
10 centimeters		3 GigaHertz
1 meter		300 MegaHertz
10 meters	Radio	30 MegaHertz
100 meters		3 MegaHertz

Figure 2.3.
The electromagnetic spectrum from wavelengths of 1 picometer to 100 meters, showing the wavelengths, frequencies, and radiation types.

than those for the color violet, is *ultraviolet* radiation; and immediately be-low visible radiation, with wavelengths longer than those for the color red, is *infrared* radiation. The wavelength scale in Figure 2.3 is presented such that each wavelength listed is 10 times the wavelength above it; for example, 1 centimeter = 10 millimeters and 1 millimeter = 1000 micro-meters. Using exponential notation, 1 centimeter = 10^{-2} meters = 0.01 me-ters, 1 millimeter = 10^{-3} meters = 0.001 meters, 1 micrometer = 10^{-6} meters, 1 nanometer = 10^{-9} meters, and 1 picometer = 10^{-12} meters. More com-pletely, the meanings of the prefixes used here and later in the chapter are listed in Table 2.1, in exponential notation, along with their standard symbols.

The boundaries indicated in Figure 2.3 between radiation types are approximate. In reality, overlaps exist, for instance between microwave and radio wavelengths and between X-rays and ultraviolet wavelengths. Furthermore, each radiation category could be subdivided into additional categories. Ultraviolet (UV) radiation includes subcategories of UV-C ra-diation for ultraviolet wavelengths shorter than 280 nanometers, UV-B ra-diation for wavelengths between 280 and 315 nanometers, and UV-A ra-diation for wavelengths between 315 and 380 nanometers. Visible

Table 2.1
Scientific prefixes and symbols in Système International (SI) units

Prefix	Meaning	Symbol	Prefix	Meaning	Symbol
deci-	10^{-1}	d	deca-	10^{1}	da
centi-	10^{-2}	c	hecto-	10^{2}	h
milli-	10^{-3}	m	kilo-	10^{3}	k
micro-	10^{-6}	μ	mega-	10^{6}	M
nano-	10^{-9}	n	giga-	10^{9}	G
pico-	10^{-12}	p	tera-	10^{12}	T
femto-	10^{-15}	f	peta-	10^{15}	P
atto-	10^{-18}	a	exa-	10^{18}	E

radiation is typically divided into the colors of the rainbow, as in Figure 2.2. Infrared radiation is typically divided into near-infrared radiation at the shortest of the infrared wavelengths (nearest the visible red radiation), thermal-infrared radiation at the middle-infrared wavelengths (about 3–15 micrometers), and far-infrared radiation at the longest of the infrared wavelengths.

Also included in Figure 2.3 is a column for the *frequency* of the radiation, a concept closely related to wavelength. This concept can be most readily explained through providing an example. For instance, if radiation has a wavelength of 1 kilometer (= 1000 meters), then, since the speed it travels through space or air is 300,000 kilometers per second, each second 300,000 wavelengths of the radiation will pass a stationary point in its path. The frequency of the radiation is consequently given as 300,000 wavelengths (or cycles) per second, and is written as 300,000 per second or, equivalently, as 300,000 s^{-1}. More generally, if the wavelength of radiation is labeled λ and the frequency is labeled f, then either can be calculated from the other using the equation

$$\lambda f = 300{,}000 \text{ kilometers per second}$$
$$= c, \text{ the speed of light.}$$

For example, at a wavelength of 1.55 centimeters, the wavelength of the microwave radiation recorded by one of the instruments on the Nimbus 5 satellite in the 1970s, the frequency of the radiation is

$$f = c/\lambda = (300{,}000 \text{ kilometers per second})/(1.55 \text{ centimeters})$$
$$= (300{,}000 \times 10^3 \text{ meters per second})/(1.55 \times 10^{-2} \text{ meters})$$
$$= 19.35 \times 10^9 \text{ per second.}$$

A common unit used to measure frequency is the Hertz, which equals 1 cycle per second, abbreviated 1 s^{-1}. Referring back to Table 2.1, a Giga-Hertz is 10^9 Hertz = 10^9 s^{-1}; thus, the frequency of radiation with a wavelength of 1.55 centimeters is 19.35 GigaHertz. Notice from the equation $\lambda f = c$ that as the wavelength increases, the frequency decreases. In Figure 2.3, the frequencies corresponding to each of the listed wavelengths are provided in the right-hand column. Because of the $\lambda f = c$ relationship and because each wavelength listed in Figure 2.3 is 10 times the wavelength above it, each frequency listed is one-tenth of the frequency above it. The prefixes, defined in Table 2.1, yield the following equivalencies: 1 MegaHertz = 10^6 Hertz = one million Hertz, 1 GigaHertz = 10^9 Hertz = one billion Hertz, 1 TeraHertz = 10^{12} Hertz, 1 PetaHertz = 10^{15} Hertz, and 1 ExaHertz = 10^{18} Hertz.

Review Questions

1. Rank the following types of radiation in order of wavelength, beginning with the radiation having the shortest wavelength: microwave radiation, ultraviolet radiation, radio-band radiation, X-rays, infrared radiation, red visible radiation, violet visible radiation.

2. What type of radiation is "near-infrared" radiation near?

3. If radiation type A has a longer wavelength than radiation type B, which one has the higher frequency?

4. What is the frequency of radiation that has a wavelength of 3.1 centimeters?

5. What is the wavelength of radiation that has a frequency of 10 ExaHertz?

Blackbody Radiation Curves

Objects generally give off radiation at an infinite number of wavelengths but much more at some wavelengths than at others. One of the primary determiners of the amount and types of radiation that an object gives off is its temperature. In fact, for every temperature, there is a precise limit as to the rate at which radiation can be given off by a body at that temperature, both for the total radiation and for each individual wavelength. A higher temperature always yields a higher limiting rate. An object that gives off radiation at the maximum rate possible at each wavelength and temperature is termed a *blackbody*. In general, actual objects are not blackbodies, but the radiation curve for the theoretical blackbody at the same temperature as the object of interest still provides a valuable reference.

Figure 2.4 shows the radiation curves for blackbodies at temperatures of 380 kelvins, 500 kelvins, and 620 kelvins. [A kelvin is a unit of temperature, generally labeled K. The numerical value of the temperature in kelvins, $T(K)$, equals the numerical value of the temperature in degrees centigrade, $T(°C)$, plus 273.15; i.e., $T(K) = T(°C) + 273.15$. Hence, 380 K = 106.85°C; 500 K = 226.85°C; 620 K = 346.85°C; and 273.15 K = 0°C, the freezing point of pure water. A key aspect of the kelvin scale is that its zero point, at -273.15°C, is theoretically the lowest temperature possible, often referred to as *absolute zero*.] Notice in Figure 2.4 that, for any two of the curves, the hotter blackbody gives off radiation at a higher rate at every positive-valued wavelength than the colder blackbody does. Still, each of the three curves (and, in fact, every blackbody radiation curve) has the

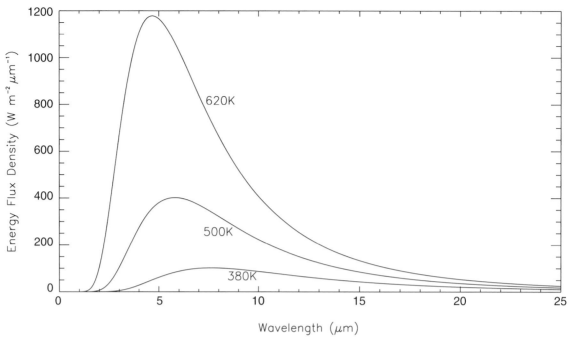

Figure 2.4.
Blackbody radiation curves for objects at temperatures of 380 kelvins (K), 500 K, and 620 K. The vertical scale is in units of watts per square meter of area, per micrometer of wavelength. One watt equals one joule per second, a joule being a standard unit of energy.

same basic shape, starting at zero radiation for a wavelength of zero, then rising as the wavelengths begin to increase, then reaching a peak, and finally dropping off and tailing toward zero radiation as the wavelengths increase further. In comparing any two different blackbody curves, the curve for the hotter object everywhere lies above the curve for the colder object (except where both are at zero) and always peaks at a shorter wavelength than the curve for the colder object (e.g., Figure 2.4).

Neither the Sun nor the Earth is a blackbody, but the blackbody curves for objects at their respective average temperatures give a rough overall indication of their radiation emissions. The Sun, with its average surface temperature of about 6000 K, is much hotter than the blackbodies whose curves are plotted in Figure 2.4. Hence, its radiation peaks at a noticeably shorter wavelength, namely, at approximately 0.5 micrometers, and at a far higher radiative emission value, in fact at more than 99,000,000 watts per square meter per micrometer. In contrast, the Earth, with an average surface temperature of approximately 285 K, is colder than the blackbodies whose radiation curves are plotted in Figure 2.4, and therefore its radiation curve peaks at a longer wavelength, approximately 10 micrometers, and at a lower emission value, approximately 24 watts per square meter per micrometer. The peak in the Sun's curve comes at a wavelength within the region of visible radiation, whereas the peak in the Earth's curve comes at a wavelength within the region of thermal-infrared radiation.

Although the Sun gives off more radiation than the Earth at every wavelength, its radiation does not dominate radiation in the Earth–atmosphere system at every wavelength. The Sun is so far from the Earth (and its radiation is spread out approximately uniformly in all directions) that the amount of radiation from the Sun that actually reaches the Earth at any wavelength is a minute fraction of the radiation given off at that wavelength. Considerable solar radiation reaches the Earth at wavelengths between 0.15 micrometers and 3 micrometers, in the vicinity of the Sun's emission peak, but very little radiation reaches the Earth from the Sun at wavelengths longer than about 7 micrometers. As a result, the natural radiation in the Earth–atmosphere system comes almost entirely from the Sun for wavelengths shorter than about 3 micrometers and almost entirely from the Earth–atmosphere system for wavelengths longer than about 7 micrometers. Sizable amounts of radiation come from both sources only in the 3–7 micrometer interval in between. This distribution is important for satellite measurements, one reason being that because of it, satellite instruments measuring at wavelengths less than about 3 micrometers require sunlight, whereas those measuring at longer wavelengths can make measurements day or night, irrespective of whether sunlight is available.

Another factor of crucial importance for satellite remote sensing of the Earth–atmosphere system is that different surfaces on the Earth and different gases and other constituents in the atmosphere emit different types and amounts of radiation. These differences allow scientists, through analysis of the radiation received by the satellite instruments, to determine something about the Earth and atmosphere from which the radiation has come. The examples in the next six chapters should help foster a sense of how this is being accomplished.

Review Questions

1. What is a *blackbody*?

2. Does any real object have a temperature whose value in kelvins is negative?

3. Water boils at a temperature of 100°C. What is this temperature in units of kelvins?

4. If blackbody radiation curves were drawn for an object having a temperature of 700 K and an object having a temperature of 900 K:

a. Which of the two curves would show higher radiation values?

b. Which of the two curves would have its peak at the longer wavelength?

5. Using Figures 2.3 and 2.4, in what region of the electromagnetic spectrum does a blackbody at a temperature of 500 K have its radiation peak?

6. Even though the Sun gives out much more radiation than the Earth at every wavelength, still the microwave radiation recorded by Earth–observing satellites comes predominantly from the Earth–atmosphere system rather than from the Sun. Why?

7. Is the visible radiation recorded by Earth–observing satellites radiation that was emitted predominantly by the Earth or by the Sun?

8. Would midnight satellite measurements of natural features in Europe in September be more effective using an instrument recording at ultraviolet wavelengths or at thermal-infrared wavelengths?

CHAPTER 3

Atmospheric Ozone and the Antarctic Ozone Hole

Introduction

Most of the Earth's atmosphere is composed of nitrogen and oxygen, but small amounts (or "trace" amounts) of other gases are also present, such as water vapor, argon, carbon dioxide, neon, helium, methane, carbon monoxide, and ozone. The presence and concentrations of some of these trace gases have important implications because of their various interactions with the radiation coming into and leaving the Earth–atmosphere system. Ozone is one of the more important trace gases in the atmosphere, and changes in its atmospheric concentrations brought about by human activities have led to considerable publicity about atmospheric ozone since the mid-1980s.

Ozone, while not itself a chemical element, consists entirely of the element oxygen. The oxygen in ozone, however, is arranged differently from the oxygen in oxygen gas, the gas vital to animal respiration (or breathing). Specifically, a molecule of ozone contains three oxygen atoms linked together, whereas a molecule of oxygen contains two oxygen atoms (Figure 3.1). This structure is reflected in the chemical symbols for the two gases: ozone is labeled O_3, and oxygen O_2.

Much of the ozone in the lower atmosphere is undesired, being one of many low-level atmospheric pollutants. Hence, reducing the amount of low-level atmospheric ozone would be beneficial. The situation is quite different in the upper atmosphere, however, where ozone performs a vital function by absorbing ultraviolet radiation from the Sun. This absorption occurs predominantly in a very important and desirable layer of air con-

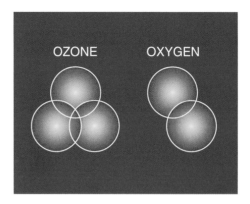

Figure 3.1.
Schematic illustration of an ozone molecule, consisting of three oxygen atoms, and an oxygen molecule, consisting of two oxygen atoms.

taining a much heavier concentration of ozone than exists in the rest of the atmosphere. Being within the region of the atmosphere termed the *stratosphere*, this layer with high ozone amounts is called the *stratospheric ozone layer*. The stratosphere lies immediately above the lowest major layer of the atmosphere, the *troposphere*, and extends from altitudes of about 9–17 kilometers to altitudes of about 45–60 kilometers, depending on location and time. Figure 3.2 shows schematically the approximate locations of the troposphere, the stratosphere, and the ozone layer. Ozone forms in the stratosphere largely because of the action of sunlight on oxygen, as radiation from the Sun (specifically, the UV-C category of ultraviolet radiation) breaks molecules of oxygen (O_2) into separate oxygen atoms (O), which then combine individually with oxygen molecules to form ozone (O_3) (Figure 3.3). The ozone, in turn, absorbs not only some of the UV-C radiation but also most of the Sun's UV-B radiation reaching the Earth's stratosphere.

Stratospheric ozone is extremely important to life at the Earth's surface. Most significantly, by absorbing ultraviolet radiation it helps protect life at the surface from the Sun's ultraviolet rays, which can cause sunburn, skin cancer, damage to the eyes, increased immune deficiencies, and other problems for humans and other plant and animal life forms.

In the early 1970s, concern arose that planned fleets of supersonic jets might alter the stratospheric ozone layer. Most of these jets were never built, and thus the ozone layer appeared safe. By the mid-1970s, however, a new concern had arisen, namely, that the chlorofluorocarbons (CFCs) entering the atmosphere as a result of their use by humans in such products as refrigerators, air conditioners, aerosol sprays, heat pumps, solvents, styrofoam insulation, and other plastic foams might precipitate a sequence of chemical reactions that would lead to stratospheric ozone re-

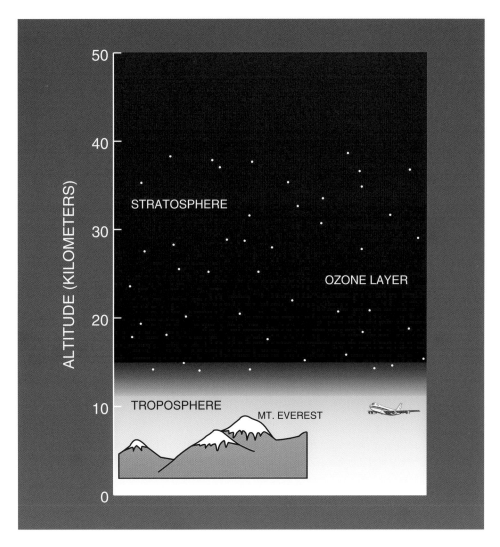

Figure 3.2.
Approximate heights in the atmosphere of Mount Everest, a cruising commercial jet aircraft, the troposphere (light blue), the stratosphere (dark blue), and the ozone layer (white dots). Satellites orbit the Earth at much higher altitudes than anything shown in this figure, generally at altitudes of at least several hundred kilometers.

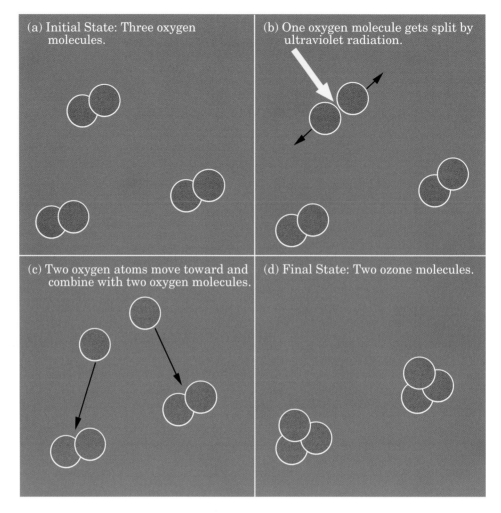

Figure 3.3.
Formation of ozone through the splitting apart of an oxygen molecule and the subsequent uniting of an oxygen atom with an oxygen molecule.

ductions. The concern, first published in 1974 by Mario Molina and F. Sherwood Rowland, was ignored by most people until a decade later, when data from the south polar region showed changes that dramatically increased public awareness of a potential problem.

In the mid-1980s, the amount of ozone in the stratosphere above the south polar region in October was found to have been decreasing over the period of the previous several years, through analysis of data taken from the British ground station at Halley Bay, Antarctica. The discovery was reported in 1985 by J. C. Farman, B. G. Gardiner, and J. D. Shanklin, and it was quickly confirmed by examination of available satellite observations. The satellite data also provided a much more complete spatial picture of how widespread the ozone decreases were. Ground-based and aircraft-based research expeditions soon followed, along with further analysis and calculations, and these established that the ozone decreases were largely caused by humans. In particular, the CFCs released into the atmosphere by human activities cause chemical reactions in the stratosphere that deplete the ozone layer, as do the halons (chemicals containing the element bromine) released into the atmosphere from fire extinguishers. As the protective ozone layer deteriorates, higher, potentially dangerous levels of ultraviolet radiation can be expected to reach the Earth's surface.

The ozone decreases in the high-latitude Southern Hemisphere stratosphere have been referred to as the *Antarctic ozone hole*. Despite the name, there is no actual hole, as some ozone is always present. The ozone decreases, however, have led to markedly low ozone amounts in the high southern latitudes during October, a Southern Hemisphere springtime month, while the middle southern latitudes from 30°S to 60°S have retained high ozone amounts. Other important factors regarding the ozone hole are that it varies from year to year (i.e., it shows *interannual variability*) and that it is not consistently becoming bigger and deeper. For instance, although the October 1987 Antarctic ozone amounts were lower than any recorded previously, the values rebounded to much higher levels in 1988, before decreasing again in 1989, to values nearly as low as those in 1987. Overall, the trend from the late 1970s to the mid-1990s has been toward a more severe springtime ozone hole over the Antarctic, but fluctuations have occurred along the way.

Subsequent to the discovery of the ozone decreases over Antarctica, decreases have been observed in stratospheric ozone levels elsewhere around the globe as well, including the Arctic and mid-latitude regions of the Northern Hemisphere. The largest decreases remain, however, in the stratosphere over Antarctica, during the Antarctic spring. A primary reason for the decreases being greatest over Antarctica centers on the formation of clouds in the extremely cold air of the Antarctic stratosphere. These stratospheric clouds contain ice particles that facilitate the chemical reactions destroying the ozone. Further, the stratosphere above Antarctica is fairly isolated (although not entirely so) during the winter and spring, with little exchange taking place with air masses from lower latitudes. Thus, the chemical changes in the Antarctic stratosphere facilitated by the stratospheric clouds are not quickly diluted either by air moving in from elsewhere or by the altered Antarctic air moving outward.

Fortunately, the environmental dangers of CFCs and halons became widely recognized in the 1980s, and in September 1987 a major event in international cooperation occurred when 27 countries signed a formal agreement to limit the further production of ozone-depleting chemicals. Signed in Montreal, the agreement is known as the Montreal Protocol.

Subsequent meetings in London in June 1990 and in November 1992 led to a strengthening of the Montreal Protocol, with phase-out of the production of CFCs, halons, carbon tetrachloride (CCl_4), and methyl chloroform (CH_3CCl_3) to be completed by the start of 1996. These changes were very welcome by many members of the scientific and environmental communities and should serve to lessen the further depletion of the ozone layer.

Review Questions

1. What two chemical elements make up most of the Earth's atmosphere?

2. **a.** What chemical element constitutes ozone?
   b. How many atoms of that element make up one molecule of ozone?

3. Is most of the ozone in the lower atmosphere (in the air we breathe) desired?

4. Why is stratospheric ozone important to life at the Earth's surface?

5. Name two problems that can be caused by excess ultraviolet radiation.

6. What human activities are contributing to stratospheric ozone loss?

7. How were the decreases in stratospheric ozone over Antarctica, later referred to as the Antarctic ozone hole, first discovered?

8. Which month of the year is highlighted here (and elsewhere) in discussion of the Antarctic ozone hole?

9. Have decreases in stratospheric ozone been observed in other locations as well as over Antarctica?

10. What constituent of stratospheric clouds facilitates the chemical reactions that destroy ozone?

11. What is the Montreal Protocol?

Satellite Detection of Ozone

A direct method of measuring stratospheric ozone would be to take samples of stratospheric air and analyze them chemically. This can be done from specially equipped aircraft or balloons, although such measurements are always limited to the specific path and time of flight of the instrument. With satellites, a much larger area can be examined and over a much longer time, but the measurements are made very differently. In particular, being done remotely, the satellite measurements cannot include direct sampling of the air. Instead, the ozone amounts are determined from the satellite data using a combination of measured radiation levels and theory.

Atmospheric ozone amounts have been calculated from data collected by the Total Ozone Mapping Spectrometer (TOMS) on board NASA's Nimbus 7 satellite, launched on October 24, 1978, and by several instruments on board the Upper Atmosphere Research Satellite (UARS), launched on September 12, 1991. Among the UARS instruments are ones with the complicated but scientifically informative names of the Cryogenic Limb Array Etalon Spectrometer (CLAES) and the Halogen Occultation Experiment (HALOE). Both the TOMS and the UARS instruments point toward the Earth and measure sunlight that has been scattered back toward space (i.e., "backscattered") from the atmosphere or the Earth's surface, after arriving from the Sun. In the case of the TOMS instrument, data from which are used in the next section to illustrate the ozone results, the measurements are made at six wavelengths in the ultraviolet region.

The basic principle on which the ozone calculations from the TOMS instrument are based is that if all other conditions remain the same, then the greater the amount of ozone in the atmosphere, the less will be the amount of ultraviolet radiation that reaches the satellite sensor. This relationship arises because of the absorption of the ultraviolet radiation by the ozone, illustrated schematically in Figure 3.4. Complications exist, however, because other atmospheric constituents also absorb some of the ultraviolet rays. If the satellite measurements were made at only one wavelength, only very rough estimates of the amount of ozone in the atmosphere could be made from the satellite data, as it would be impossible

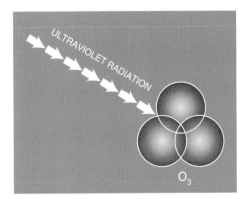

Figure 3.4.
Schematic illustration of the absorption of ultraviolet radiation by ozone, a process central to the theory behind the calculation of ozone amounts from TOMS satellite data.

to sort out how much absorption of the ultraviolet rays was due to the presence of ozone and how much was due to the presence of other atmospheric constituents. The fact that the TOMS instrument makes measurements at several wavelengths helps those analyzing the data to sort out the ozone contributions and increase the accuracy of the calculated ozone amounts. Of particular importance in this regard, the fraction of ultraviolet radiation that gets absorbed by ozone differs for the six wavelengths measured by the TOMS, and the pattern of differences is not identical to the patterns produced by other atmospheric constituents. All this contributes to a complicated atmosphere but helps in the calculation of how much ozone must be in the atmosphere to account for the measured radiation values at each of the wavelengths. This calculation of atmospheric ozone levels is done with an *algorithm* (or set of equations) for converting from the radiation measurements to the ozone amounts.

The ozone values obtained from the TOMS instrument are generally presented in "Dobson units," named after G. M. B. Dobson for his early work in the 1920s on the stratospheric ozone layer and its connection to stratospheric air circulations. Dobson units measure the thickness of the layer of pure ozone that would be created if all the ozone in the atmosphere were isolated from the rest of the atmosphere and brought down to sea level, adjusting to the much greater atmospheric pressure at sea level. A single Dobson unit corresponds to a thickness of 0.01 millimeters, or, equivalently, 0.001 centimeters. Hence, a somewhat typical atmospheric ozone measurement of 300 Dobson units corresponds to a thickness of only 3 millimeters (about one-tenth of an inch); i.e., ozone in the atmosphere, however important, is very sparse.

The amounts of ozone calculated from the TOMS data are thought to be very good approximations to the actual amounts of ozone in the atmosphere. They are not considered perfect, however, as the algorithms used for the calculations do not incorporate all of the many chemical components of the atmosphere or all of their potential interactions with the Sun's ultraviolet radiation. Clouds can also complicate the interpretation of the data, as can the details of the distribution of the ozone vertically through the atmosphere.

Because the TOMS instrument measures reflected sunlight, its data cannot be used for determining ozone amounts during periods of darkness. In the equatorial regions, this simply means that the measurements need to be made during the day rather than at night; but in the polar regions, where in winter the Sun does not rise above the horizon at all for weeks or months at a time, the result is extensive periods without data. The extreme case arises precisely at the north and south poles, where the Sun is not above the horizon at any time during the entire six months of autumn and winter, then remains continually above the horizon for the six months of spring and summer. The situation is less extreme further equatorward but exists to some extent throughout the region from either pole equatorward to 66.5° latitude. The effect lessens as the distance from the pole increases, until at 66.5° latitude there is only one day (the summer solstice) during which the Sun is above the horizon for the entire 24 hours, that is, only one day with a "midnight Sun," and correspondingly only one day (the winter solstice) during which the Sun is continually below the horizon.

Review Questions

1. Do satellite instruments collect air samples, thereby allowing the direct measurement of ozone?

2. Which satellite instruments have been used for ozone calculations?

3. Briefly, how are ozone amounts determined from satellite observations?

4. An algorithm could be constructed concerning almost any topic and could consist simply of a single equation. What is an appropriate algorithm for converting from the number of hours (H) in a time interval to the number of minutes (M) in that same interval?

5. What units are used to measure atmospheric ozone?

6. If all the ozone in the atmosphere were brought vertically down to sea level and adjusted to the atmospheric pressure at sea level, into which of the following ranges would the thickness of the resulting ozone layer fall: 0 to 1 centimeter, 1 centimeter to 1 meter, 1 meter to 1 kilometer, greater than 1 kilometer?

7. How might other chemical components of the atmosphere affect the accuracy of the ozone amounts calculated from the TOMS satellite data?

8. Can the TOMS measurements be used to determine polar ozone amounts throughout the year? If not, when are the measurements not useful?

Satellite Images of Ozone

A global map of average atmospheric ozone amounts, in Dobson units, for the month of March 1990 is presented in Figure 3.5a. The figure is color-coded so that low ozone amounts are depicted in shades of blue and purple, and high ozone amounts are depicted in shades of yellow and orange. The scale is set to cover a range from a low of 115 Dobson units to a high of 500 Dobson units. There are times and locations where values lower than 115 Dobson units are obtained by the TOMS instrument, and there are also times and locations where values higher than 500 Dobson units are obtained, but on average over the month March 1990, no such unusually low or high values occurred, so that the scale used depicts the full range of average March 1990 values. The color coding incorporates 11 colors, each covering a range of 35 Dobson units.

Figures 3.5b, 3.6a, and 3.6b present maps of average atmospheric ozone amounts for the months of June, September, and December 1990, all using the same color scale as in Figure 3.5a. The four images show marked seasonal differences in ozone amounts, differences that are expanded upon in the questions at the end of the section. It should be noted that whereas in March and September the TOMS instrument could record ozone amounts globally, it could not do so in June and December, because of the "polar nights." In June, the month of the Southern Hemisphere's winter solstice, when the south polar region is in darkness throughout the day and night, the instrument was unable to collect ozone data south of about 66.5°S; hence, that region is left black in the June 1990 image of Figure 3.5. Similarly, in December, the month of the Northern Hemisphere's winter solstice, the north polar region is in darkness, preventing the TOMS instrument from collecting ozone data north of about 66.5°N and resulting in the north polar region being colored black on Figure 3.6b.

Figure 3.7 shows the development and relaxation of the Antarctic ozone hole in 1990, as depicted on monthly average images for August through November. The black area in the center of the August image signifies missing data resulting from the lack of sunlight in the Southern Hemisphere winter. It is a continuation of the polar night that reached its maximum spatial extent in June (Figure 3.5b) at the time of the Southern Hemisphere's winter solstice. The September image in Figure 3.7 depicts the same data as shown for the Southern Hemisphere in Figure 3.6a, although with a very different projection, one centered on the south pole. The reader should compare the two images sufficiently to become convinced that both indeed display the same Southern Hemisphere data (and do so using the same color scale). The polar projection gives a much better indication of the shapes of Antarctica and of the ozone hole above Antarctica, although it fails to capture the global picture because of not depicting the Northern Hemisphere data.

Figure 3.8 illustrates interannual variations in ozone amounts and distributions by showing four October images for the south polar region, again using the same color scale as in Figures 3.5–3.7. Of these years, the ozone hole was most prominent in 1985, being both more intense and centered closer to the south pole than in 1979, 1982, or 1988. Although the hole weakened between 1985 and 1988, it deepened again in 1990, as revealed by comparison between Figure 3.8 and the October image of Figure 3.7. The interannual variability shown among the October ozone images

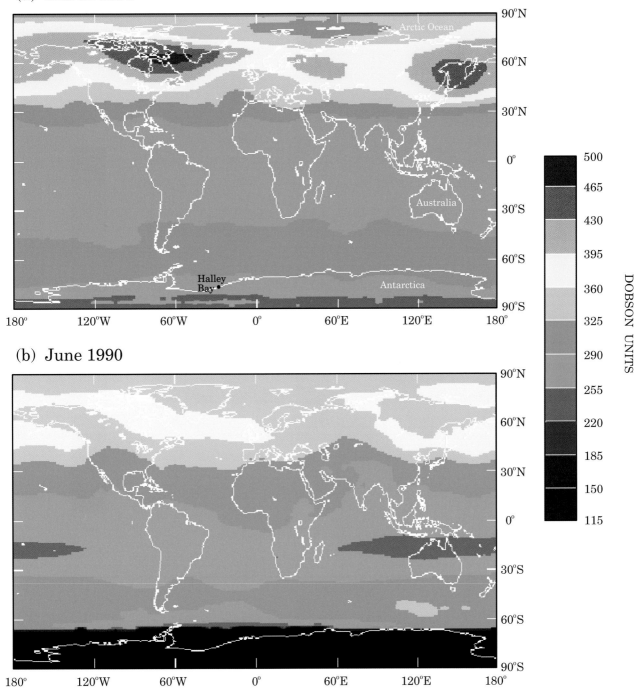

(a) March 1990

(b) June 1990

Figure 3.5.
(a) Average global ozone amounts for March 1990, calculated from the data of the Nimbus 7 TOMS. (b) Average global ozone amounts for June 1990, calculated from the data of the Nimbus 7 TOMS. Both images were obtained on CD-ROM from the TOMS Ozone Processing Team at NASA Goddard Space Flight Center, with labels added later.

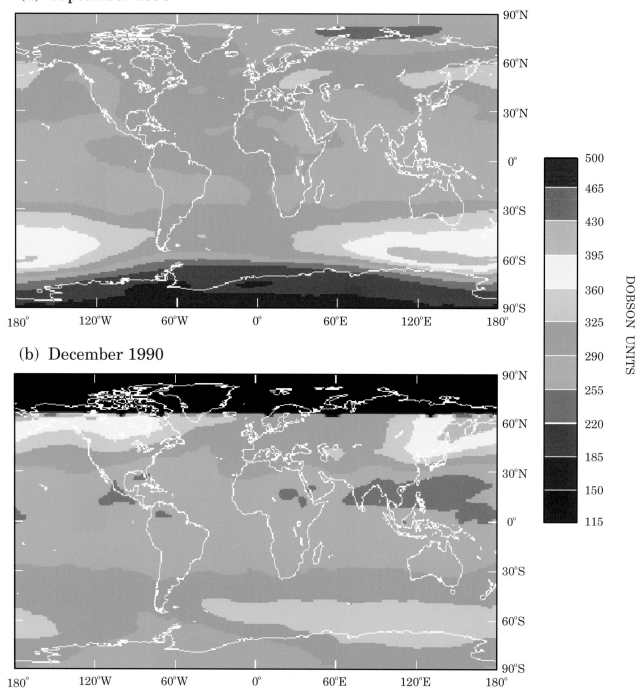

(a) September 1990

(b) December 1990

DOBSON UNITS

Figure 3.6.
(a) Average global ozone amounts for September 1990, calculated from the data of the Nimbus 7 TOMS. (b) Average global ozone amounts for December 1990, calculated from the data of the Nimbus 7 TOMS. Both images were obtained on CD-ROM from the TOMS Ozone Processing Team at NASA Goddard Space Flight Center, with labels added later.

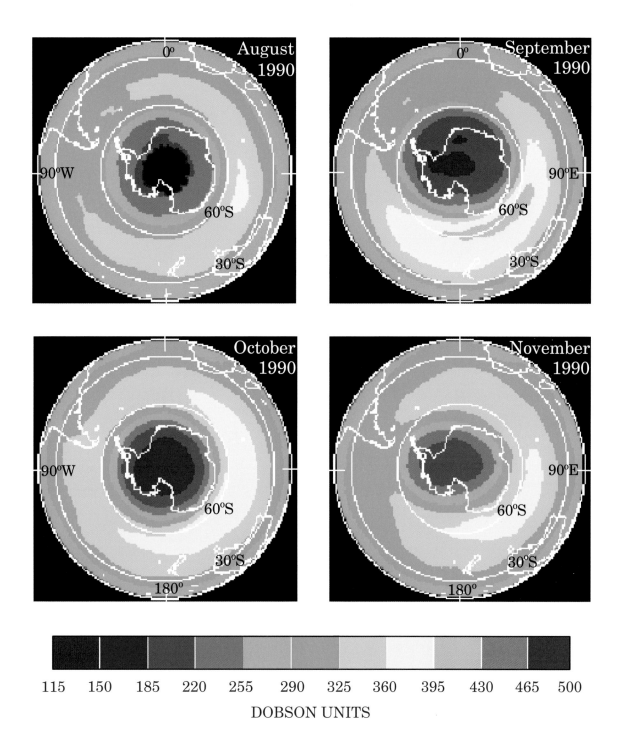

115   150   185   220   255   290   325   360   395   430   465   500

DOBSON UNITS

Figure 3.7.
Average Southern Hemisphere ozone amounts for August, September, October, and November 1990, calculated from the data of the Nimbus 7 TOMS. The
images were obtained on CD-ROM from the TOMS Ozone Processing Team at
NASA Goddard Space Flight Center, with labels added later.

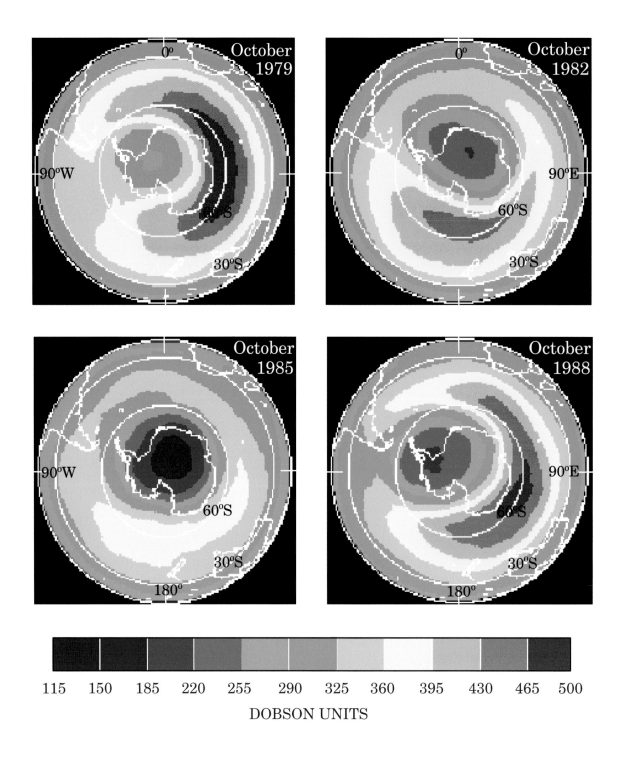

Figure 3.8.
Average Southern Hemisphere ozone amounts for four Octobers (1979, 1982, 1985, and 1988), calculated from the data of the Nimbus 7 TOMS. The images were obtained on CD-ROM from the TOMS Ozone Processing Team at NASA Goddard Space Flight Center, with labels added later.

30

of Figures 3.7 and 3.8 is considerable, both in terms of the spatial patterns and in terms of the amounts of ozone. Atmospheric ozone varies from year to year to a much greater degree than some of the variables depicted in later chapters, including sea ice and continental snow coverage in Chapters 4 and 5.

The ozone images of Figures 3.5–3.8 were all obtained from the CD-ROM (Compact Disk–Read Only Memory) titled "TOMS Ozone Image Data 1978–1991" put out by Pat Guimaraes of Hughes STX and Rich McPeters of the NASA Goddard Space Flight Center. These data are available through the Global Change Data Center, NASA Goddard Space Flight Center, Greenbelt, Maryland.

Questions Regarding the Satellite Imagery

Note: In all questions in this section, "ozone" refers to atmospheric ozone.

1. According to Figure 3.5a, which hemisphere had the highest ozone amounts in March 1990?

2. Where were the highest ozone amounts, on average, in March 1990, according to Figure 3.5a, and what were their values in Dobson units?

3. Where were the lowest ozone amounts, on average, in March 1990, and what were their values in Dobson units?

4. Where were the highest ozone amounts, on average, in September 1990, and what were their values in Dobson units?

5. Where were the lowest ozone amounts, on average, in September 1990, and what were their values in Dobson units?

6. Order the months March, June, and September 1990 according to the amount of ozone in the Northern Hemisphere, starting with the month of most ozone.

7. Order the months March, September, and December 1990 according to the amount of ozone in the mid-latitudes (30–60°S) of the Southern Hemisphere, starting with the month of most ozone.

8. Explain the similarities in the answers to Questions 6 and 7.

9. **a.** In which of the months March, June, September, and December 1990 were the ozone amounts greater in the equatorial regions than in the mid-latitudes of the Northern Hemisphere?

   b. In which of these months were the ozone amounts greater in the equatorial regions than in the mid-latitudes of the Southern Hemisphere?

10. On a monthly average basis, which month, of August, September, October, and November, in 1990 had the highest ozone amounts in the mid-latitude regions between Australia and Antarctica?

11. On a monthly average basis, which month, of August, September, October, and November, in 1990 had the largest region of very low ozone amounts (less than 185 Dobson units) over Antarctica?

12. a. What are the highest ozone amounts indicated in Figure 3.8 for October 1979, and where are they located?
    b. What are the highest amounts indicated for October 1982?
    c. What are the highest amounts indicated for October 1985?
    d. What are the highest amounts indicated for October 1988?

13. a. What are the lowest ozone amounts indicated in Figure 3.8 for October 1979?
    b. What are the lowest amounts indicated for October 1982?
    c. What are the lowest amounts indicated for October 1985?
    d. What are the lowest amounts indicated for October 1988?

14. In which of the years 1979, 1982, 1985, and 1988 was the October ozone hole centered closest to the south pole?

CHAPTER 4

Polar Sea Ice

Introduction

Sea ice is ice formed through the freezing of sea water. Because of the cold temperatures in the polar regions and the large ocean area, sea ice spreads over vast expanses of the polar seas. The ice takes many forms, from individual ice crystals suspended in the water to sheets of ice extending unbroken for several kilometers. A thin sheet of ice is unlikely to remain unbroken for long, except under extremely calm conditions, but as the ice thickens it becomes more able to withstand disruptive pressures, produced by such forces as winds, waves, currents, and tides, and remain intact for months, years, or even decades. A sizable sea ice cover is typically broken into ice floes ranging from several centimeters to several meters thick and from tens of centimeters to several kilometers across. The floes tend to be interspersed with open-water "leads" through the ice, with the leads in between thick ice floes often containing younger, thinner ice forms within them. All the ice pieces are subject to deformation and movement when acted upon by winds, water, and other influences, such as ships maneuvering in the vicinity. Figure 4.1 presents photographs of a sampling of different sea ice types, all of which float on the water and in general have about 90% of their mass and volume below the level of the sea surface.

Sea ice differs considerably from icebergs, the other major category of ice floating in polar seas. The most fundamental difference is that sea ice is formed in the sea by the freezing of water, whereas icebergs are formed by the breaking off into the sea of glacier ice originally formed on land from the accumulation and solidification of fallen snow. Icebergs can be gigantic, hundreds of meters thick, whereas sea ice is generally less than

(a)

(b)

(c)

(d)

(e)

(f)

Figure 4.1.
Sea ice of different forms and from different perspectives. (a) Sparsely distributed ice floes, as viewed from a ship. (b) Expansive ice field, as viewed from an aircraft. (c) Close-up of newly formed ice. (d) Ice floes separated by a lead, as viewed from an aircraft. (e) Thin sheets of ice, as viewed from an aircraft. (f) Several-months-old ice bearing the weight of a helicopter, as viewed from ground level. [Figure 4.1e was kindly provided courtesy of Koni Steffen; the other photographs are from the author.]

5 meters thick (except where ice floes have been pressed together to form ridges, which can rise well above the original ice level, and keels, which extend deep beneath the ridges, the ridge–keel combination sometimes reaching thicknesses of 20 meters or more, still well below the thicknesses of typical icebergs). Also, because sea ice forms from sea water, it tends to be salty, whereas icebergs consist of fresh water. Hence, the ice from icebergs is much more appropriate than sea ice for use as ice cubes or for melting and then drinking—an important distinction for individuals attempting to survive on the polar seas. However, although sea ice is salty, it is generally not as salty as sea water, the reason being that much of the salt content does not freeze but instead drops to the water underneath. This release of salt from the ice occurs both during the freezing process and later, after some of the salt has migrated downward through the ice floe.

The release of salt from the ice to the underlying water can have important impacts on ocean circulation. The added salt increases the density of the water directly under the ice, and this can be particularly important because, at the low temperatures found in the polar regions, variations of water density occur primarily because of variations in salinity, unlike the situation in low latitudes, where variations in temperature tend to dominate. In polar oceans, the increase in density caused by the release of salt sometimes suffices to make the water directly under the ice denser than the water further down. When that happens, an overturning occurs, as schematized in Figure 4.2a. This overturning can precipitate additional, larger-scale changes in ocean circulation both near the surface and further down. When the overturning transports water from the surface layers to the bottom levels of the ocean, *deep water* has been formed. In fact, much of the deep water in the Earth's oceans is believed to have "formed" (i.e., descended to become deep water, having already been water but closer to the surface) in the polar and subpolar regions in the vicinity of the sea ice cover. Some of this deep-water formation occurs through the aforementioned consequences of salt rejection as ice forms and ages, and some occurs through the consequences of the very different process of water densification through heat loss to the atmosphere. The latter process is significant in the northern North Atlantic, where salty water moves poleward, cools, sinks, and becomes part of the North Atlantic deep water.

The impact that sea ice has on water circulation is only one of several impacts of sea ice on the non-sea-ice components of the Earth's climate system. Another important impact results from the role of sea ice as an insulator, restricting the transfers of heat, mass, and momentum between the atmosphere and the ocean. For instance, in the middle of winter, when the air temperatures are generally considerably lower than the water temperatures, which do not descend much below the freezing point, heat can be transferred from the ocean to the atmosphere at rates

(a) Salt Release

(b) Insulation

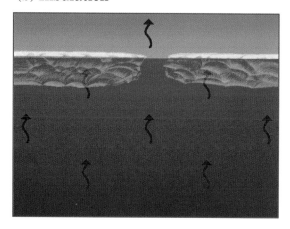

(c) Reflectivity

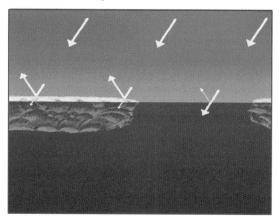

(d) Non–Climate Impacts

Figure 4.2.
Schematic views of some of the major climate and nonclimate impacts of sea ice. (a) The release of salt (white dots in the diagram) by the ice to the underlying ocean increases the density of the water and sometimes leads to sinking and overturning. (b) The ice serves as an effective insulator, reducing the transfers of heat (red arrows) and other variables between the ocean and the atmosphere. (c) In marked contrast to open water, sea ice reflects most of the solar radiation (yellow arrows) incident on it, lessening the amount of radiation absorbed at the surface. (d) The ice can be a danger to ships but also a platform for humans and other animal life.

exceeding 500 watts for every square meter. Such rates occur, however, only where the water has direct contact with the atmosphere (and in fact with a very cold atmosphere), for instance in the leads or other spaces between ice floes. An intervening layer of ice greatly lessens the vertical heat transfers (Figure 4.2b). The thicker the ice, the more effective it is as an insulator, with ice of thickness one meter or greater generally reducing the heat flows to less than 25 watts per square meter. The ice also reduces the exchanges of mass, for instance mass in the form of water vapor, between the ocean and the atmosphere, and it absorbs some of the momentum (calculated as mass times velocity) transferred by the winds. The absorption of momentum leaves less momentum to be transferred to wave motions in the underlying waters, an effect much appreciated by many crew members and passengers on ships in polar waters during windy conditions. Ice-resistant ships can often escape the rough waters just outside the ice edge by moving into the ice pack, where the waters are relatively calm because of the reduced momentum transfer, providing welcome relief to anyone subject to seasickness.

Another important effect of the ice results from its tendency to reflect a large percentage of the Sun's radiation incident on it. Visually, sea ice appears white or near white, in great contrast to the ocean water around the ice and between the ice floes. The ocean water appears dark because most of the Sun's radiation that strikes it gets absorbed into it, whereas the ice appears white because most of the Sun's radiation that strikes the ice gets reflected (Figure 4.2c). The contrast in absorption between the water and the ice means that the presence of ice on the polar oceans reduces the amount of solar radiation absorbed at the surface. If the ice cover decreases in extent, either seasonally or from year to year, then the oceans will absorb more energy and can be expected to heat up. The effect of the contrasting reflectivities between ice and water is enhanced when a snow cover overlies the ice—a common occurrence in both polar regions—because snow generally reflects an even larger percentage (80%–97% for new snow) of the incident solar radiation than bare ice does (closer to 50%, versus approximately 5%–15% for ocean water).

Sea ice has many other impacts as well, in addition to those on the climate system. It can be a major obstacle and hazard to ships in its vicinity (Figure 4.2d); it greatly complicates the interpretation of submarine acoustics; it hinders oil exploration in polar waters, although it can also be used as a platform for oil drilling; and it is a platform regularly used by polar wildlife, including polar bears in the Northern Hemisphere and penguins in the Southern Hemisphere (e.g., Figure 4.2d).

All the various impacts of the ice would have little overall consequence if sea ice were rare. In actuality, however, sea ice extends over a substantial area, mostly in the polar regions. At any given time, sea ice covers approximately 25 million square kilometers of the Earth's oceans, making it more extensive than the entire North American continent (24.4 million square kilometers). In the Northern Hemisphere, sea ice coverage is greatest in March, at the end of winter, and least in September, at the end of summer, extending over about 15 million square kilometers of the Northern Hemisphere in March and about 8 million square kilometers of the Northern Hemisphere in September. In the Southern Hemisphere, where the seasons are reversed from those in the Northern Hemisphere, sea ice extent is generally greatest in September, at about 19 million square kilometers, and least in February, at about 4 million square kilometers. In both cases, the reason that sea ice coverage is greatest at the end of winter rather than earlier is that even though the temperatures have warmed up from the previous month, they are still cold enough at the end of winter to allow additional ice to form and to prevent most of the existing ice from melting. Satellite images later in the chapter show geographically where the ice is.

As scientists contemplate changes in global climate, sea ice is one of the many variables being examined. Climatically important for the past many thousands of years because of its large areal coverage and its influence on the atmosphere and oceans, sea ice in the past two decades has gained added potential in the field of climate change research because of how readily the global sea ice cover can now be monitored, through satellite observations. Until the 1970s, obtaining an accurate picture of global sea ice coverage was almost impossible, but now, with current satellite instruments, large-scale sea ice coverage is readily ascertained on a routine basis, something still not possible for most other climate variables. Thus, for instance, if the wintertime conditions in the north polar region were to become significantly harsher, this change would likely become apparent earlier from satellite data showing increased sea ice coverage than it would from measurements of atmospheric temperatures.

Review Questions

1. a. What is sea ice?
   b. Is the thickness of sea ice generally greater than or less than the height of a two-story house?

2. What is the fundamental difference between sea ice and icebergs?

3. Place the following in order of saltiness, from least salty to most salty: sea water, icebergs, sea ice.

4. Why does the saltiness of sea ice differ from that of sea water?

5. By what mechanism does ice formation sometimes contribute to deep-water formation?

6. Does the presence of ice increase or decrease the amount of heat transferred between the ocean and the atmosphere?

7. If the polar regions warm and the sea ice cover retreats, what will be the effect on the amount of solar radiation absorbed at the surface of the polar oceans?

8. Is the area covered by sea ice around the globe generally closer to the area of Japan (370,000 square kilometers), the area of the United States (9.5 million square kilometers), the area of North America (24.4 million square kilometers), or the area of the Pacific Ocean (181.3 million square kilometers)?

9. During a typical Arctic winter, do the coldest overall temperatures tend to come before, after, or at the same time as the greatest sea ice areal coverage?

10. What modern technology now enables global sea ice coverage to be monitored on a fairly routine basis?

Satellite Detection of Sea Ice

Sea ice can be detected much more easily from space than can many other climate variables. For instance, it can be detected with visible radiation (e.g., Figure 1.4), in sharp contrast to variables such as ozone and carbon dioxide, which cannot be distinguished from the rest of the atmosphere by the human eye or by instruments recording at visible wavelengths. Two major problems with visible radiation for detecting sea ice, however, have already been mentioned in Chapter 1: (1) clouds present between the ice and the atmosphere obscure the view of the ice cover; and (2) visible images require sunlight and hence cannot be obtained during the polar night.

A different type of satellite instrument that avoids these two complications and that can also be used to detect sea ice is a passive-microwave radiometer. A *radiometer* is an instrument that measures radiation; *microwave* refers to the type of radiation being measured (see Figure 2.3); and *passive* refers to the fact that the instrument simply receives the radiation from elsewhere (i.e., is passively making the measurements). *Active* instruments, in contrast to passive ones, actually send out a signal that they later receive back. Sea ice has been examined with satellite active-microwave radiometers as well as satellite passive-microwave radiometers and with satellite instruments recording at infrared wavelengths, the latter having the drawback that the sea ice cover is often obscured by overlying clouds. However, the samples in this chapter are each from passive-microwave radiometers.

Three primary satellite passive-microwave radiometers that have provided a wealth of sea ice information are the Nimbus 5 Electrically Scanning Microwave Radiometer (ESMR), which provided data for most of the period from its launch in December 1972 until the end of 1976; the Nimbus 7 Scanning Multichannel Microwave Radiometer (SMMR), which provided data for most of the period from its launch in October 1978 until the middle of August 1987; and a series of Special Sensor Microwave Imagers (SSMIs) on the satellites of the Defense Meteorological Satellite Program (DMSP), which have provided data for most of the period since June 1987. The ESMR was a single-channel instrument, collecting data at a wavelength of 1.55 centimeters, whereas the SMMR was a ten-channel instrument, collecting both horizontally polarized and vertically polarized data at each of five wavelengths, and the SSMIs are seven-channel instruments, collecting both horizontally polarized and vertically polarized data at three wavelengths plus vertically polarized data at a fourth wavelength. The multichannel nature of the SMMR and SSMIs offers the possibility of examining more variables and obtaining greater accuracies than is possible with a single channel of information.

Passive-microwave radiometers are particularly valuable for sea ice studies because of the very sharp contrast between the emission of microwave radiation from sea ice and the emission of microwave radiation from liquid water. Even though sea ice is colder than water, at many wavelengths it emits radiation at a much higher rate than water does. Neither sea ice nor water is a blackbody, and hence neither emits radiation at as high a rate as a blackbody would at the same temperature (Chapter 2). However, because of differences in the electromagnetic properties of the two media, at many wavelengths sea ice comes much closer to emitting at a blackbody rate than does water. The ratio of the rate of emission, at any particular wavelength, of a substance to the rate at which a blackbody would emit at the same temperature and wavelength is called the *emissivity* of the substance at that wavelength. For an example, at the microwave wavelength of 1.55 centimeters, water has an emissivity of only about 0.44 but most sea ice has an emissivity somewhere between 0.80 and 0.97 (Figure 4.3). This much higher emissivity for sea ice than for water outweighs the fact that water has a higher temperature and leads to the emission rate for sea ice being much higher than for water, at the 1.55-centimeter wavelength. A similar situation arises at many other microwave wavelengths as well.

The sharp contrast in microwave emissions between ice and water, along with the fact that the transition region between open water and a fairly compact ice cover tends to be narrow, enables the ready determination of the location of the ice edge from the passive-microwave data. Examples for the wintertime ice covers of each polar region are shown in

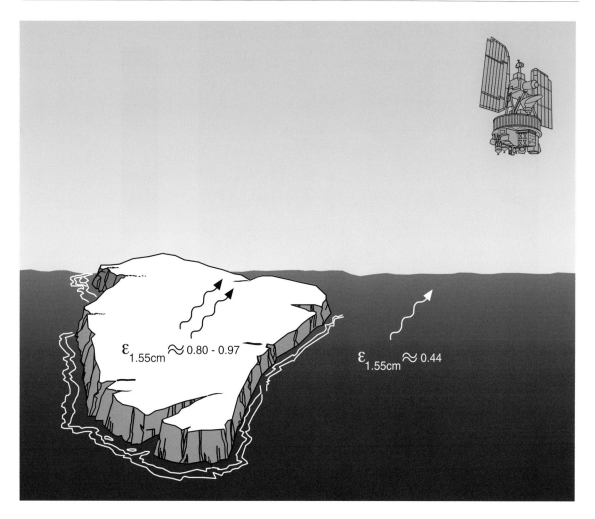

Figure 4.3.
Schematic illustration showing the higher rates of microwave emission from
sea ice than from open water, resulting from higher microwave emissivities, ϵ,
for sea ice. The emissivities indicated are those for a wavelength of 1.55
centimeters.

Figure 4.4, using data from the Nimbus 5 ESMR. These images are color-
coded in units referred to as *brightness temperatures*, which are measures
closely corresponding to the radiation values received by the satellite instrument but presented in units of temperature (generally kelvins, abbreviated K). The dashed white line showing the location of the ice edge has
been overlain to identify the approximate ice-edge location. On the open-
ocean side of the dashed line, the brightness temperatures have values
generally below 160 K, but on the ice side, the values are much higher,
generally at least 190 K. That is, the ice has much higher brightness tem-

(a) North polar region, March 8–10, 1974

(b) South polar region, September 16–18, 1974

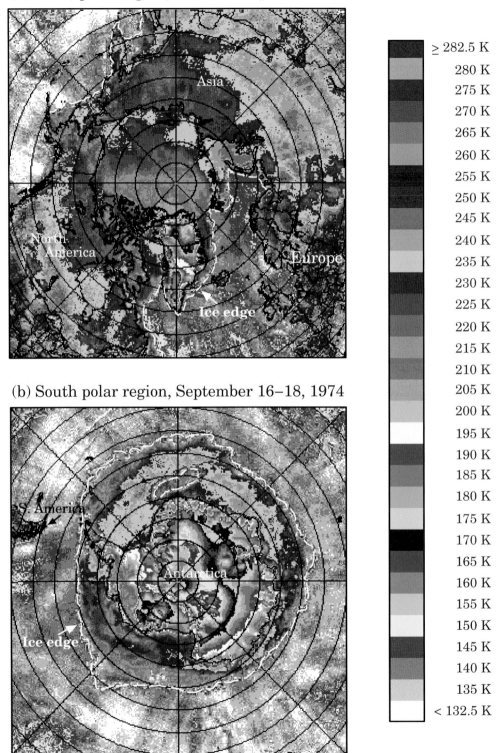

	≥ 282.5 K
	280 K
	275 K
	270 K
	265 K
	260 K
	255 K
	250 K
	245 K
	240 K
	235 K
	230 K
	225 K
	220 K
	215 K
	210 K
	205 K
	200 K
	195 K
	190 K
	185 K
	180 K
	175 K
	170 K
	165 K
	160 K
	155 K
	150 K
	145 K
	140 K
	135 K
	< 132.5 K

Figure 4.4.
Brightness temperature images of the north polar region for the three-day period March 8–10, 1974, and of the south polar region for the three-day period September 16–18, 1974, as obtained from the data of the Nimbus 5 ESMR.

42

peratures than the water, reflecting its higher rate of microwave emission. The satellite techniques would work just as well if the sea ice emitted much less microwave radiation than water, as long as the user of the data knew the basic differences; the key requirement for the ice to be readily distinguishable is simply that there be a marked difference in the recorded radiation values, ice versus water.

The passive-microwave data also provide the opportunity to calculate how compact an ice cover is. As is clear from the photographs of Figure 4.1, ice floes are massed together very compactly at some times and locations but are dispersed widely apart at other times and locations. The term used to indicate the percentage of an ocean area that is overlain by ice is *ice concentration*. A region with no ice has an ice concentration of 0%, whereas a region half covered with ice has an ice concentration of 50%, and a region fully covered with ice has an ice concentration of 100%. The ice concentration can be calculated from passive-microwave data for the same basic reason that the ice edge can be determined from the passive-microwave data, namely, because the ice and the water emit microwave radiation at markedly different rates. When a satellite instrument collects data from the surface, it does so for a fairly sizable surface area. In the case of satellite passive-microwave sea ice data, a single radiation value transmitted by the satellite is generally derived from a region with a diameter anywhere from about 12 kilometers to about 150 kilometers, depending upon the particular satellite instrument, the wavelength used, and the altitude of the satellite. If the region had a very high ice concentration at the time of the observation, then the satellite instrument would record a high radiation value typical of sea ice; but if the region had a very low ice concentration, near 0%, then the satellite instrument would record a much lower radiation value typical of water, with intermediate ice concentrations resulting in intermediate radiation values. Hence, the radiation values can be used to determine at least approximate ice concentrations.

If ice concentration were the only factor affecting the microwave radiation received by the satellite, then the ice concentrations presumably could be determined from the satellite data not just approximately, as indicated in the previous paragraph, but highly accurately. Unfortunately (in the context of the calculation of ice concentration), many additional factors also affect how much radiation is received by the satellite. These additional factors include the thickness of the ice, the salinity of the ice, whether a snow cover lies on the ice, whether any melting is occurring on the snow or the ice, and what the atmospheric conditions are. All of these factors complicate the accurate determination of ice concentration from satellite passive-microwave data. However, fortunately, none of them is as dominant as the ice concentration is in determining the satellite's receipt of microwave radiation, and so at least approximate concentrations can still be calculated, despite the complicating factors. Furthermore, when the calculations incorporate data from more than one channel of microwave information, attempts can be made to sort out some of the complications, and, as a result, the accuracies can be improved. Current estimates of the accuracies of wintertime sea ice concentrations from calculations using only one channel of microwave data, such as from the single-channel Nimbus 5 ESMR, are about ±15%, whereas those from calculations using three or four channels of microwave data, such as from the Nimbus 7 SMMR or the DMSP SSMIs, are about ±7%. The improved accuracies narrow the range of where the true ice concentrations are thought to lie. For instance, with a calculated ice concentration of 75%, an accuracy of ±15% means that the actual ice concentration probably lies between 75% – 15% = 60% and 75% + 15% = 90%, whereas an accuracy of ±7% means that the actual ice concentration probably lies between 68% (= 75% – 7%) and 82% (= 75% + 7%). In many scientific usages, *accuracy* has a much more precise meaning; but in the case of derived sea ice concentrations, the estimated accuracies remain rough. As instruments improve, the accuracies can be expected to improve also. Accuracies in summer are generally not as good as those in winter, in large part because of the uncertainties associated with surface melting, which is far more prevalent in summer than in winter.

Review Questions

1. Which of the following three variables can be seen most clearly on satellite images obtained from visible radiation data: sea ice, ozone, or carbon dioxide?

2. What is the difference between a passive-microwave instrument and an active-microwave instrument?

3. What contrast is particularly valuable for the determination of sea ice locations from satellite passive-microwave data?

4. **a.** Physically, which is warmer, ice or water?
   b. Which emits microwave radiation at a higher rate, ice or water?

5. At any particular wavelength, what is the emissivity of a blackbody?

6. Rank the following four surface types in order of how high their brightness temperatures are, in general, in Figure 4.4, from lowest to highest: the sea ice covers of the Arctic and Antarctic; the ice sheets of Greenland and Antarctica; the land surfaces of central Europe and the northern United States; the ice-free oceans.

7. If three-fourths of a region of the ocean is covered by sea ice, what is its average sea ice concentration?

8. For each of the following photographs of Figure 4.1, indicate whether the average ice concentration for the ice shown is closer to 0%, to 45%, or to 90%:
   a. Figure 4.1a.
   b. Figure 4.1b.
   c. Figure 4.1e.

9. What are some of the factors that can affect how much microwave radiation is given off by an ice floe?

Satellite Images of Sea Ice

Figures 4.5 and 4.6 show a pair of color-coded satellite images of sea ice concentrations in the Northern Hemisphere in the year 1986. The ice concentrations indicated in both images were derived from the data of the Nimbus 7 SMMR. Figure 4.5 shows the calculated ice concentrations averaged over March, the month of maximum ice coverage, and Figure 4.6 shows the calculated ice concentrations averaged over September, the month of minimum ice coverage. Correspondingly, Figure 4.7 shows a pair of color-coded satellite images of sea ice in the Southern Hemisphere in 1986, the top image showing ice concentrations for February, the month of minimum ice coverage in the Southern Hemisphere, and the bottom image showing ice concentrations for September, the month of maximum ice coverage in the Southern Hemisphere. On each of the four images, land has been artificially set to gray, as no sea ice exists on the land and the geography becomes much more apparent when all the land areas are set at the same color. The ice concentrations are color-coded into 22 shades of blue, green, yellow, orange, brown, red, and purple, representing sea ice concentrations ranging from 12% to 100%, with each color shade corresponding to an ice concentration increment of 4%. Figure 4.8 provides maps indicating the locations of the various places referred to in the text and questions.

The circular black areas extending from 84.6°N to 90°N in the images of Figures 4.5 and 4.6 indicate missing data resulting from the fact that the orbit of the Nimbus 7 satellite did not go close enough to the north pole to allow the SMMR instrument to collect data north of 84.6°N. (The orbit also did not allow data collection south of 84.6°S, but that region is part of the Antarctic continent and hence is colored gray in Figure 4.7.) Notice that this underlying reason for the black, missing-data areas of Figures 4.5 and 4.6 differs fundamentally from the reason that such areas appear in the ozone images of Figures 3.5–3.7. In the ozone case, the instrument was able to collect data globally but the lack of sunlight led to missing data during the polar night.

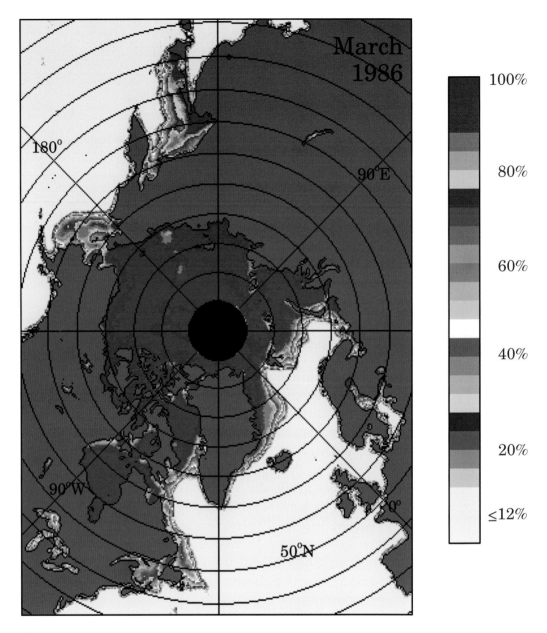

Figure 4.5.
Monthly average sea ice concentrations for the north polar region for March 1986, calculated from the data of the Nimbus 7 SMMR.

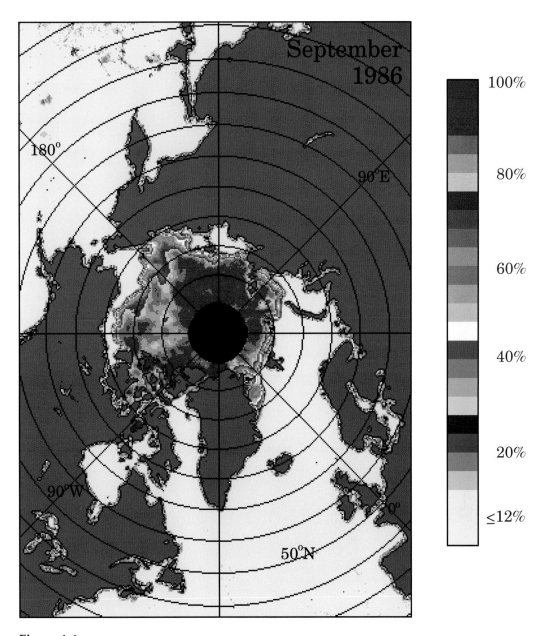

Figure 4.6.
Monthly average sea ice concentrations for the north polar region for September 1986, calculated from the data of the Nimbus 7 SMMR.

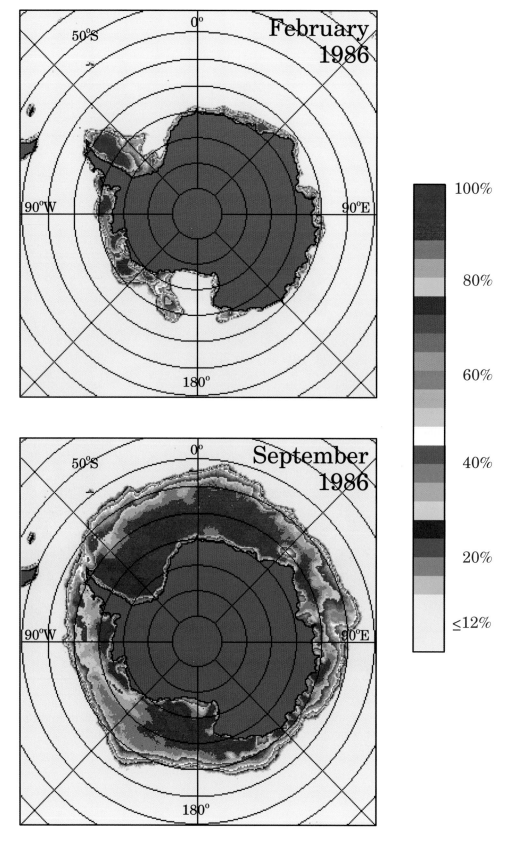

Figure 4.7.
Monthly average sea ice concentrations for the south polar region for February 1986 and September 1986, calculated from the data of the Nimbus 7 SMMR.

North Polar Region

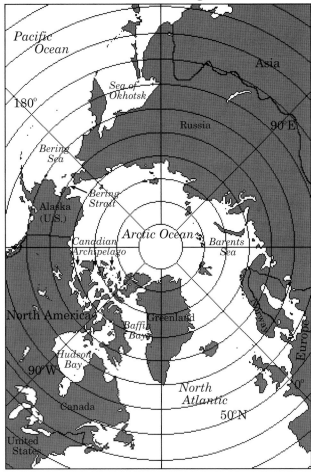

South Polar Region

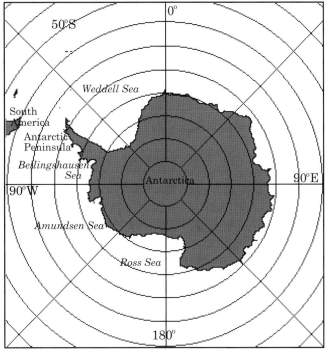

Figure 4.8.
Location maps for the north and south polar regions.

The uniformity of coloring for most of the ice-free ocean area in Figures 4.5–4.7 is partly artificial, resulting because the ocean has been given a consistent light blue color wherever the ice concentration was calculated as equal to or less than 12%. This coloring scheme removes much of the "noise" that would otherwise occur in the images at locations where the calculations indicate some sea ice even though no ice exists. Noise can be produced from a variety of causes, such as geographic variations in the ocean surface temperature or atmospheric winds roughening the surface of the ocean and causing changes in its emission of microwave radiation. Some noise can be seen in the Pacific region of the Northern Hemisphere's monthly September 1986 image (Figure 4.6). Daily images typically show more noise, but much of the atmospherically induced noise is removed through monthly averaging.

A separate cause for false indications of sea ice, and one that is not removed through monthly averaging, arises in response to a situation peculiar to land–ocean boundaries. All along these boundaries, the calculations commonly indicate ice even far from any ice-covered regions (e.g., along the coast of southwest Europe in Figures 4.5 and 4.6 and along the coast of South America in Figure 4.7). This error occurs because, for each data point, the satellite collects radiation from a large area, about 50 kilometers in diameter for the relevant SMMR data, and therefore along the coasts any individual data point generally has some radiation coming from land and some from ocean. Because land data typically have much higher brightness temperature values than ice-free ocean data (e.g., Figure 4.4) and the ice calculations are based on the fact that ice has higher values than ice-free ocean, when the calculations are performed for data points that include land as well as ocean, the high values from the land averaged in with the ocean values make the results look like ice exists in the ocean along the coast even where no ice occurs. This problem is termed *land contamination*, because the satellite data for the ocean are being contaminated by having some radiation from the land mixed in with the radiation from the ocean. For that reason, the halo of colors around the land boundaries should be ignored, as the sea ice calculations are not reliable adjacent to the land.

The appearance of atmospheric noise and land contamination on satellite sea ice images can be greatly reduced and sometimes even eliminated by various techniques. Most primitively, a person can manually clean the images, replacing by the light-blue, ice-free ocean coloring all indications of ice presence in areas where the individual is confident that no ice exists. Another, less labor-intensive scheme would be to provide the computer with maps of sea surface temperature (SST) and instructions (in the form of computer code) to set the ice concentrations to 0% wherever the SST exceeds some predetermined value (something above the freezing point). Whatever method is used, subjectivity is involved, and some risk enters that correct indications of sea ice might get masked along with the incorrect indications of ice. Still, the methods can improve both the appearance of the images and the overall accuracy of the amount of ice displayed. Neither the manual cleaning nor an SST adjustment was made here, in order to leave the images closer to what the initial calculations provide and to illustrate and discuss the various complications and cautions. (The sea ice algorithm does, however, include a *weather filter*, i.e., an empirical numerical scheme to remove much of the weather-related noise.)

The images in Figures 4.5–4.7 provide a wealth of information about the distribution and extent of sea ice in the polar regions. Answering the questions at the end of the section will help the reader identify many of the types of information that the images contain. For a sampling, notice how unsymmetrical the Northern Hemisphere sea ice distributions are in March (Figure 4.5). The warm Gulf Stream waters moving northeast across the North Atlantic help keep much of the North Atlantic free of ice year-round even at latitudes north of 65°N, not just in 1986 (Figure 4.5) but in all years of the modern record. In contrast, Hudson Bay is at much lower latitudes but is almost fully ice covered (Figure 4.5). The reason lies not only in the Bay's lack of access to warm waters from the south but also in its geographic placement in the midst of a continent. Because land surfaces heat and cool much more rapidly than do ocean surfaces subjected to the same intensity of solar radiation, continental areas tend to be warmer in summer and cooler in winter than mid-ocean areas at the same latitude, a phenomenon sometimes referred to as the *continentality effect*. Hudson Bay, being in the midst of North America, is affected by the impacts of the land around it and thus tends to be colder in winter, and more likely to have ice formed, than mid-ocean areas at the same latitudes.

The ESMR and SMMR radiative data used to obtain the images in Figures 4.4–4.7 are available through the National Snow and Ice Data Center (NSIDC) in Boulder, Colorado, as are the radiative data from the later SSMI instruments. A detailed description of the algorithm used to obtain the sea ice concentrations mapped in Figures 4.5–4.7 can be found in the book *Arctic and Antarctic Sea Ice, 1978–1987: Satellite Passive-Microwave Observations and Analysis* by P. Gloersen, W. J. Campbell, D. J. Cavalieri, J. C. Comiso, C. L. Parkinson, and H. J. Zwally, listed under "Additional Reading." Together, the SMMR and SSMI data sets provide satellite passive-microwave observations of global sea ice coverage from late October 1978 until the present. The record shows strong seasonal cycles in each hemisphere in each year, with fair consistency (at least through 1996) in the timing of maximum and minimum hemispheric ice coverage. Considerable interannual variability occurs in the ice distributions and extents, however, especially in some of the regional ice covers, such as in Baffin Bay in the Northern Hemisphere and in the Weddell Sea in the Southern

Hemisphere. Overall, the satellite passive-microwave record between 1978 and 1995 shows neither a prominent increasing trend nor a prominent decreasing trend in the ice coverage of either hemisphere.

Questions Regarding the Satellite Imagery

1. **a.** What is the cause of the missing data (colored black) in the sea ice images of Figures 4.5 and 4.6?
   b. What is the cause of the missing data (also colored black) in the ozone images of Figures 3.5–3.7?
   c. In view of the contrast between the reasons for the missing data, in which data set (ozone or sea ice) does the location of the missing data stay the same throughout the year?

2. In the color scale of Figures 4.5–4.7, what sea ice concentration is associated with each of the following color breaks:
   a. The break between yellow and green.
   b. The break between green and blue.
   c. The break between brown and pink.

3. The text mentions land contamination and atmospherically induced noise as two common reasons for false indications of ice in satellite-derived maps of sea ice concentrations.
   a. For which of these two does monthly averaging significantly reduce the problem?
   b. Why is the problem reduced for that category but not for the other category?

4. **a.** Which of the following Northern Hemisphere regions have an appreciable ice cover in late summer, as indicated in the September 1986 map of Figure 4.6: the northern Sea of Okhotsk, the northern Bering Sea, the Arctic Ocean, Hudson Bay, Baffin Bay, the Canadian Archipelago, the southern Barents Sea, the northern Barents Sea? (See Figure 4.8 for locations.)
   b. Which of the locations listed under Question 4.a have an appreciable ice cover in late winter, as indicated in the March 1986 map of Figure 4.5?

5. **a.** What were the approximate average sea ice concentrations in the central Arctic Ocean in March 1986, according to the data in Figure 4.5?
   b. What were the approximate average sea ice concentrations in the central Arctic Ocean in September 1986?

6. **a.** Of the northern Sea of Okhotsk, the Bering Sea, and Hudson Bay, which one has the highest wintertime ice concentrations, as suggested by the data from March 1986?

b. Provide a geophysical explanation of the answer to Question 6.a, recognizing that the three localities are all at approximately the same latitude.

7. From the way the sea ice is distributed in the Sea of Okhotsk in March, does it appear likely that the warm surface waters entering the Sea of Okhotsk from the Pacific do so by moving north along the eastern portion of the sea or along the western portion of the sea?

8. In September 1986, would it have been easier for a ship that had just entered the Arctic through the Bering Strait to proceed several hundred kilometers along the north coast of Alaska or several hundred kilometers along the north coast of Russia? (In contrast to the answers to most of the questions in this section, this answer would not necessarily be the same if the data were for a year other than 1986.)

9. **a.** What is the approximate latitude of the ice edge in the Barents Sea in March?
   b. What is the approximate latitude of the ice edge in the Barents Sea in September?

10. **a.** West of Norway, a strong surface current exists termed the Norwegian Current. From the March distribution of ice in this region, does it appear that the Norwegian Current flows northward or that it flows southward?
    b. Why?

11. Which sea in the Antarctic region has the most ice remaining at the end of summer, as indicated by the data from February 1986?

12. **a.** In September of 1986, at approximately what longitude did the sea ice cover of the Southern Hemisphere extend farthest to the north?
    b. How far north did it extend at that longitude?

13. What September 1986 ice concentration is indicated on Figure 4.7 for the location 90°W, 70°S?

14. Which hemisphere appears to have the largest area of very high ice concentrations, 96%–100%, at the end of winter (March in the Northern Hemisphere, September in the Southern Hemisphere), as indicated in the 1986 data?

CHAPTER 5

Continental Snow Cover

Introduction

Like sea ice, continental snow at times covers a large area of the Earth's surface (up to 40% of the land surface in the Northern Hemisphere winter) and has prominent local impacts both on the surface underlying it and on the atmosphere overlying it. Snow covers most of the continent of Antarctica and the island of Greenland (the Earth's largest island) throughout the year, plus large areas of North America, Europe, and Asia in winter. It also covers high-altitude mountain peaks throughout the world, amongst the best known being the Himalayas in Asia, the Alps in Europe, the Rockies in North America, the Andes in South America, and Mount Kenya and Mount Kilimanjaro in Africa, the latter two within 4° latitude of the equator. Snow also typically covers a large portion of the sea ice in both polar oceans, but because sea ice is highlighted in Chapter 4, this chapter concentrates on snow coverage over land.

Snow is composed of ice crystals in a porous, permeable state. It originates in the atmosphere as individual, fragile crystals that take various forms depending on the temperature and humidity at the time and place of formation. As snow descends through the atmosphere, it can further transform depending upon the ambient temperature, humidity, and wind conditions. Once on the ground, the crystals rapidly undergo further metamorphosis, and the grains become more strongly bonded together. In freshly fallen snow, the original shape of the individual crystals remains clear, whereas in older snow, the individual crystals are often no longer recognizable. The snow begins to compact soon after falling, especially as more snow falls on top of it, and continues to compact over time. Where the snow cover survives year after year, failing to melt or evaporate off during the summers, increased compaction eventually transforms the snow to firn, an intermediate stage between snow and ice, and then to ice. Densities tend to be in the ranges of 0.01–0.3 grams per cubic centimeter for new snow, 0.2–0.6 grams per cubic centimeter for old snow, 0.40–0.84 grams per cubic centimeter for firn, and 0.84–0.92 grams per cubic centimeter for glacier ice, compared with 1 gram per cubic centimeter for water. New snow has particularly low densities if the only conditions it has been subjected to are calm and very cold. No matter what the original density of a snow cover, however, the density tends to increase with time, in response to the compaction process. When the density reaches about 0.83 grams per cubic centimeter, the air spaces between the snow grains close off and the medium approaches glacier ice. This is unlikely to happen in a seasonal snow cover that melts away during the spring and summer, although seasonal snow packs do experience other transformations, occurring, for instance, whenever rain or surface melt results in water percolating through the snow.

Like sea ice, snow has both climate and nonclimate impacts (Figure 5.1). In fact, several of the climate impacts are quite similar to those for sea ice, and for similar reasons. Snow, like ice, reflects most of the solar radiation that reaches it (Figure 5.1a). Its solar reflectivity (or *albedo*) is generally even greater than that of ice, as a layer of new snow typically reflects between 80% and 97% of the visible solar radiation incident on it. Older, dirtier snow reflects less, sometimes under 40%, but this reflectivity still often significantly exceeds that of the surface under it. Even in wooded areas, where trees sticking through the snow cover absorb radiation at a much higher rate than the snow and thereby reduce the areally averaged reflectivity, the presence of the snow still keeps the reflectivity higher than it otherwise would be. Typical reflectivities for snow-covered forests are 25%–40%, versus about 9%–18% for forests without a snow cover. Thus, in forested and nonforested areas alike, the presence of a snow cover decreases the solar radiation absorbed at the surface. Because most of the reflected radiation passes through the atmosphere and returns to space but the absorbed radiation remains available for heating, the reduced absorption due to the presence of snow has a cooling effect. During the melt season, the retreat of the snow cover can be greatly accelerated once a few snow-free patches of ground appear, as the higher radiation absorption in the snow-free patches makes more energy available to melt the surrounding snow cover. Similarly, on the much longer time frame of ice ages, once deglaciation begins, the process is aided as the ice and snow retreat, exposing much-less-reflective surfaces and thereby resulting in the absorption of more energy, some of which subsequently becomes available for further melting.

A second climate impact of snow, again matching a corresponding impact of sea ice, results from its insulating properties (Figure 5.1b). Snow,

(a) Reflectivity

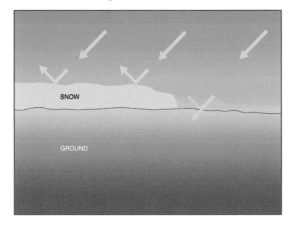

(b) Insulation

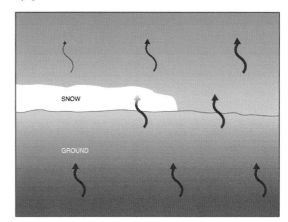

(c) Transportation Impediment

(d) Recreational Opportunities

Figure 5.1.
Illustrative views of some of the major climate and nonclimate impacts of snow.
(a) Snow has a very high reflectivity, resulting in less solar radiation being absorbed at the surface in the presence of a snow cover than in its absence. (b)
Snow provides insulation between the ground and the atmosphere, significantly restricting heat exchanges. (c) Amongst its detrimental nonclimate impacts, snow impedes conventional means of transportation. (d) Amongst its
beneficial nonclimate impacts, snow makes possible numerous recreational activities. [Photographs in c and d are from the author.]

like ice, does not conduct heat well (i.e., it has a low *thermal conductivity*), so that heat from the ground is not lost as rapidly to the cold winter atmosphere when a snow cover exists. This is particularly true when the snow cover consists of fresh snow, which has a much lower thermal conductivity than does denser, older snow. The insulation provided by a snow cover reduces frost penetration into the ground.

Another climate impact of a seasonal snow cover relates to the "thermal inertia" it contributes to the local climate system, for instance slowing temperature increases during the spring and summer melt season. As the snow melts, a solid (snow) is converted to a liquid (water), a transformation that requires energy. In fact, 334,000 joules of energy are required to melt 1 kilogram of snow. The consumption of energy during the melt process leaves less energy for warming the atmosphere, thereby slowing the melt-season temperature increases.

Snow also has a host of nonclimate impacts, some harmful and some beneficial. First, snow can be a major hazard, as evidenced in avalanches, deathly blizzards, and treacherous driving conditions produced even from a very slight snowfall. During a snowstorm and for the subsequent few days, schools, government agencies, and commercial activities can all be brought to a near-standstill locally and regionally on the basis of snow clogging the transportation arteries (Figure 5.1c). How much snow is required for such effects depends greatly on how well prepared the particular locality is. Many high-latitude communities can handle snowfalls of almost any magnitude, but lower-latitude cities unused to heavy snowfalls, such as Washington, D.C., can become nearly crippled by snow amounts that elsewhere would be considered fairly minor or routine. Even for localities well prepared, however, the economic costs of snow removal necessary to maintain normal activities can be substantial.

Snow avalanches can bring death not only to mountain climbers and skiers caught in their midst but also, in severe cases, to individuals well away from the mountain slopes. For instance, in 1962 a massive avalanche of ice and snow from Mount Huascarán in the Peruvian Andes mixed with soil and rock as it descended the lower slopes, then traveled 16 kilometers into the Santa Valley, destroying nine towns and killing 4000 people. This disaster was highly unusual, but it illustrates the deathly force an avalanche can have under extreme circumstances.

Despite the hazards, snow can also be vastly beneficial, as evidenced in the beauty of "winter-wonderland" scenes, the popular recreational activities made possible by snow, such as cross-country skiing, downhill skiing, or snow sculpting (Figure 5.1d), and the life-sustaining fresh water provided when snow melts. For many communities in or near snow-covered mountainous areas, a large part of the drinking water comes from reservoirs fed by springtime snow melt. Snow melt is also used for hydropower and for water for irrigation. Where the water is not siphoned off for these and other purposes, the snow melt generally is deposited in one or more of the following ways: it enters the groundwater reservoir; it moves downhill, often in streams or rivers, eventually settling in lakes or oceans; or it is evaporated back into the atmosphere. Typically, 10 centimeters of snow reduces when melted to about 1 centimeter of water (over the same area), although this equivalence can vary significantly, from about 30 centimeters of very loose snow melting down to 1 centimeter of water to about 2 centimeters of old, wet snow melting down to 1 centimeter of water. The amount of water created as the snow melts can be critical for communities dependent on the melt for drinking water or other uses, or for communities in danger of flooding from excessive or rapid snow melt.

Regarding the seasonal cycle of snow coverage, in the Northern Hemisphere seasonal snowfall in general begins in mid-September, typically with the first snow falling in Siberia (northeastern Russia) and northern Alaska. However, unusual September snowfalls can occur almost anywhere in northern North America or northern Eurasia (Europe and Asia). Snow coverage in the far north expands and thickens in October and November, contributing, as it expands, to further cooling, because of its high reflectivity. By mid-December, snow coverage generally includes much of Alaska, Canada, Russia, and the Scandinavian countries of Norway, Sweden, and Finland. Snow extent increases through the winter months, generally reaching a maximum sometime in February, with a hemispheric extent typically of about 45 million square kilometers. Retreat of the snow cover proceeds basically northward from March through June. By mid-June, the seasonal snow is confined mostly to the far north, along the Arctic coasts. At the summer minimum, the areal extent of the Northern Hemisphere land snow cover is about 4 million square kilometers. Later in the chapter, sample satellite images provide maps of the Northern Hemisphere snow cover distribution for February, March, April, and May 1986.

In the Southern Hemisphere, snow generally covers more than 90% of Antarctica throughout the year. Comparatively speaking, even in winter not much snow exists elsewhere in the Southern Hemisphere, due in large part to how little additional land area in the Southern Hemisphere is located at high latitudes. This is only comparative to the vast expanse of snow on Antarctica, however, as the Southern Hemisphere does have locally expansive snow covers on several major mountain peaks and mountain ranges. Even in the equatorial regions, at high enough altitudes snow can be found in abundance (Figure 5.2). Still, in terms of the total Southern Hemisphere snow coverage, Antarctica dominates, and, as its areal coverage of the continent remains greater than 90% year round, the result is a far smaller seasonal contrast in total hemispheric snow coverage for the Southern Hemisphere than for the Northern Hemisphere. The

Figure 5.2.
Snow coverage at a latitude of 9°S, at Pucahirca, Peru, in the northern Andes, October 1991. The altitude in the foreground, where a satellite-linked weather station is being prepared, is 5325 meters. [Photograph kindly provided courtesy of Lonnie Thompson.]

Southern Hemisphere areal snow coverage ranges from a minimum of about 13 million square kilometers near the end of summer to a maximum of about 16 million square kilometers near the end of winter.

Review Questions

1. Which continent is largely snow-covered throughout the year?

2. Name three atmospheric variables that affect snow as it forms and descends through the atmosphere.

3. Place the following in order of average densities, from least to most dense: water, old snow, new snow, glacier ice, firn.

4. Place the following in order of typical solar reflectivities, from least to most reflective: old snow, new snow, snow-covered forests, forests without a snow cover.

5. How does snow's solar reflectivity affect the air temperatures in the vicinity of the snow?

6. What effect does the fact that snow provides insulation between the ground and the atmosphere have on frost penetration into the ground in the midst of winter?

7. **a.** Name some of the ways that snow can be a hazard.
   b. Name some of the ways that snow can be beneficial.

8. Consider two snow covers, each of the same area and thickness but one consisting of new snow and the other consisting of old snow. If both of these snow covers were to be completely melted down, which one would probably produce the most melt water?

9. In general, when does seasonal snow coverage begin in the Northern Hemisphere, and where?

10. In which month does snow coverage in the Northern Hemisphere generally reach its annual maximum?

11. Place the following in order of the area of snow coverage in a typical year, from lowest to highest: summer minimum snow coverage in the Northern Hemisphere; winter maximum snow coverage in the Northern Hemisphere; summer minimum snow coverage in the Southern Hemisphere; winter maximum snow coverage in the Southern Hemisphere.

Satellite Detection of Snow

Northern Hemisphere continental snow cover has been monitored since 1966 using visible and infrared satellite instruments. The United States National Oceanic and Atmospheric Administration (NOAA), formed in 1970, has been a primary participant in this activity and has compiled weekly snow cover data sets from the visible and infrared instruments going back to 1966. The data from these instruments are particularly good at delimiting the snow cover for freshly fallen, highly reflective snow, and the NOAA data sets have been used in many studies examining the seasonal and interannual changes in the snow cover. Two limitations of the NOAA visible and infrared data sets, however, are that: (1) the snow data are obscured whenever a cloud cover lies overhead, and (2) the data sets do not provide estimates of snow thickness and thus cannot be used to obtain volume estimates relevant to fresh water supplies and potential flood situations. Passive-microwave data, in contrast, although not extending as far back in time as the visible and infrared data, are generally not obscured by cloud cover and have been used to calculate estimates of snow thicknesses as well as snow coverages and extents. In fact, the passive-microwave data from the same SMMR and SSMI instruments highlighted in Chapter 4 regarding sea ice determinations are also used for continental-snow calculations. All the snow-cover results presented in the next section are from the SMMR passive-microwave data set.

The basic principle by which satellite passive-microwave data have been used to obtain estimates of snow cover thickness over land surfaces centers on the fact that much of the radiation received by the satellite from a moderate snow cover overlying a land surface actually emanates from the land underneath the snow, not from the snow itself. As the radiation travels upward through the snow, it gets scattered by the snow crystals and grains, reducing the amount that reaches the satellite instrument. The thicker the snow cover, the more scattering occurs and the smaller the amount of radiation that emerges from the top of the snow cover (Figure 5.3). Thus, a rough snow-thickness algorithm could be based on a single channel of microwave information: the less radiation recorded by the satellite, the greater the estimated snow thickness.

Several complicating factors hinder the determination of snow thickness through rough algorithms using only one data channel. In particular: different ground surfaces emit different amounts of radiation, so that no single value is uniformly appropriate for the no-snow case; the amount of radiation emitted by any surface changes as the surface temperature changes; the amount of scattering of radiation within the snow depends upon the density of the snow cover and the sizes of the snow grains and crystals, not just the snow thickness; the radiation received by the satellite includes emission from the snow as well as from the land underneath; the

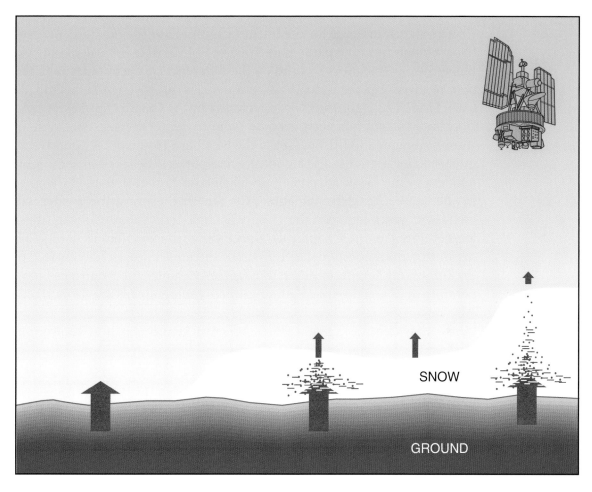

Figure 5.3.
Schematic illustration showing the scattering within a snow cover of microwave emissions from the ground and hence the tendency for thicker snow covers to result in lower microwave brightness temperatures.

microwave characteristics of snow change markedly under conditions of melt; and buildings, trees, and other objects often rise well above the snow surface, with emissions from these also contributing to the radiation received by the satellite. All of these complications limit the accuracy that can be expected when snow thicknesses are calculated from a single channel of microwave information. They generally remain complications when the calculations are done using more than one channel, but by judicious use of additional channels, the uncertainties caused by at least some of the complications can be reduced and the accuracies can often be improved. Consequently, many algorithms for obtaining snow thickness estimates from the microwave data use two or more channels of information.

The algorithm used for the images presented in the next section uses two channels from the passive-microwave SMMR data set, calculating estimated snow thicknesses based on the difference in the amount of radiation received by the satellite at frequencies of 18 GigaHertz and 37 GigaHertz. The algorithm was developed by Al Chang, Jim Foster, and Dorothy Hall and has been used by them to depict the seasonal cycle of snow coverages and thicknesses in the Northern Hemisphere, plus many interannual variations. The algorithm incorporates the fact that snow scatters less of the lower frequency, 18 GigaHertz radiation than the higher frequency, 37 GigaHertz radiation. The thicker the snow, the greater will be the difference between the 18 and 37 GigaHertz values (or brightness temperatures, as in Chapter 4). Specifically, the calculation sets snow depth in centimeters equal to 1.59 times the difference, in kelvins (K), of the brightness temperature at 18 GigaHertz minus the brightness temperature at 37 GigaHertz. (Both brightness temperatures are for horizontally polarized data.) The formulation was derived for dry snow conditions, snow densities of 0.30 grams per cubic centimeter, and snow grain sizes of 0.35 millimeters, although it is applied under a much broader range of conditions. The complications mentioned in the previous paragraph still exist, but the algorithm has been found to provide better results overall than when either the 18 GigaHertz or 37 GigaHertz data are used alone.

The snow-thickness algorithm of Chang, Foster, and Hall applies only for moderate snow depths (about 5–70 centimeters) and only over non-ice-covered land surfaces. It does not apply over ice sheets, such as the Greenland ice sheet, where a snow cover exists almost all the time but the results of the algorithm generally indicate no snow. The reason the algorithm fails to detect the Greenland snow is that a basic assumption in the algorithm—that much of the microwave radiation reaching the satellite instrument comes from the land underneath the snow—is no longer valid. In the case of Greenland, the radiation from the underlying land is almost entirely scattered away within the great thickness of the ice sheet.

Over the 1978–1987 period of the SMMR record, wintertime snow cover extents in the Northern Hemisphere calculated from the SMMR data varied by about 12%, with snow coverage least in 1981 and greatest in 1985. Snow extents over the same time period have also been calculated with other satellite data, in particular the NOAA visible and infrared data. In general, the estimated snow extents calculated from the SMMR data are about 10% less than those calculated from the NOAA data. The difference is believed to stem largely from the fact that the SMMR algorithm does not reveal snow shallower than about 5 centimeters, whereas the NOAA maps show the shallow snow as well as the thicker snow. Both techniques reveal the same basic patterns of change in snow extents within a year and from year to year, but the difference in the numerical values obtained underscores the need, whenever trends or other changes in the snow cover are being examined, to use a consistent data set throughout the period being analyzed.

Review Questions

1. What are two major limitations of the NOAA visible and infrared satellite data for snow studies?

2. What is an advantage of the NOAA visible and infrared satellite data over the satellite passive-microwave data for snow studies?

3. For a moderate snow cover over land, why is the amount of microwave radiation received by the satellite instrument generally less as the snow cover gets thicker?

4. Several factors are listed in the text that complicate calculations of snow thickness from a single channel of microwave radiation. Which two of these complications make it impossible to identify a perfect land-surface radiative value for use in the algorithm, from which to subtract radiation scattered within the snow?

5. Why can the difference between the radiative values recorded by the satellite instrument at frequencies of 18 GigaHertz and 37 GigaHertz (both for horizontally polarized data) be used to calculate estimated snow thicknesses?

6. Why does the passive-microwave snow-thickness algorithm described not apply over most of Greenland?

7. **a.** Are the snow extents calculated using the satellite passive-microwave data generally greater than or less than the snow extents calculated using the NOAA visible and infrared data?
   b. Why?

8. Assume a situation in which snow coverage over North America was the same in each of two consecutive years. Assume further that a person is comparing values for snow extents calculated for the first year from NOAA visible and infrared data and for the second year from passive-microwave data, not knowing that the values came from different instruments. Which of the following three conclusions would this person most likely reach: (a) snow coverage was the same in the two years, (b) snow coverage increased from the first year to the second year, or (c) snow coverage decreased from the first year to the second year?

Satellite Images of Snow Coverage

Figures 5.4–5.7 show a sequence of monthly average images of continental snow coverage in the high latitudes of the Northern Hemisphere for the year 1986. The first two images are for February and March, two of the three months (January through March) of greatest snow coverage in the Northern Hemisphere; and the next two images are for April and May, during the period of rapid poleward retreat of the snow cover. The same color scale is used in each image, to allow ready comparisons amongst the months. In this scale, snow thicknesses of 4 centimeters or less are ignored and are colored the same gray color as areas without snow. For snow thicknesses between 4 centimeters and 70 centimeters, separate color shades are used for each thickness interval of 3 centimeters. All calculations were done using the data of the Nimbus 7 SMMR and the algorithm described in the last section, basing the snow thicknesses on the difference between the brightness temperatures of two channels in the microwave data. Figure 5.8 provides a location map for these images and the questions at the end of the section.

The most glaring problem in Figures 5.4–5.7 is the lack of snow coverage indicated over Greenland. As discussed in the previous section, Greenland is almost entirely snow-covered, but the surface under the snow is an ice surface rather than a land surface, and as a result, the radiative processes are dramatically different and the algorithm used for calculating snow thicknesses over land is not appropriate. Hence, the values calculated for Greenland and other ice-covered areas (such as portions of Iceland and Svalbard) should be ignored.

Just as land was set to a uniform gray in the sea ice images of Chapter 4, because land has no sea ice on it, so here for the continental snow images, all ocean areas are set to a uniform blue, because ocean has no continental snow.

To illustrate some of the differences in snow coverage between different years, Figures 5.9 and 5.10 present monthly average February images for 1979 and 1981. The overall average snow coverages in these two Februarys, as well as that in February 1986 shown in Figure 5.4, are very similar, although interannual differences are clearly visible, as they are between the monthly averages for any selected month and any two years of the satellite record. Several of the prominent differences are highlighted in the questions at the end of this section. Overall, these snow-cover differences, however, are nowhere near as striking as the interannual differences in atmospheric ozone amounts, as illustrated, for instance, in Figure 3.8, where major differences appear not just in the amounts of atmospheric ozone but also in the spatial patterns of the ozone distributions. Northern Hemisphere snow coverage, although variable from year to year, is considerably less variable than atmospheric ozone.

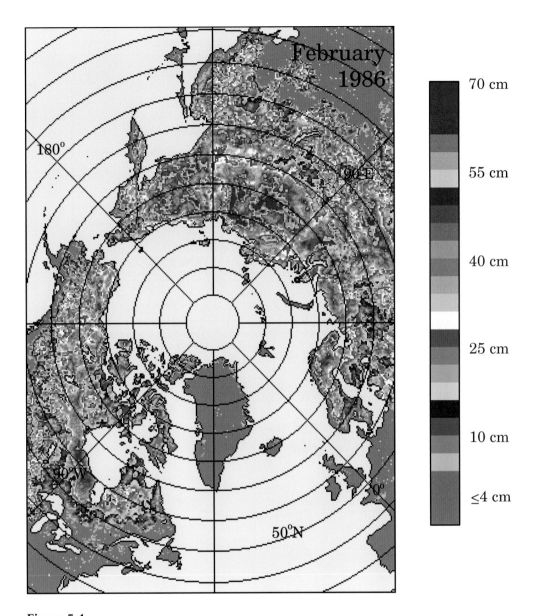

Figure 5.4.
Monthly average snow thicknesses over continental land surfaces in the north polar region for February 1986, calculated from the data of the Nimbus 7 SMMR.

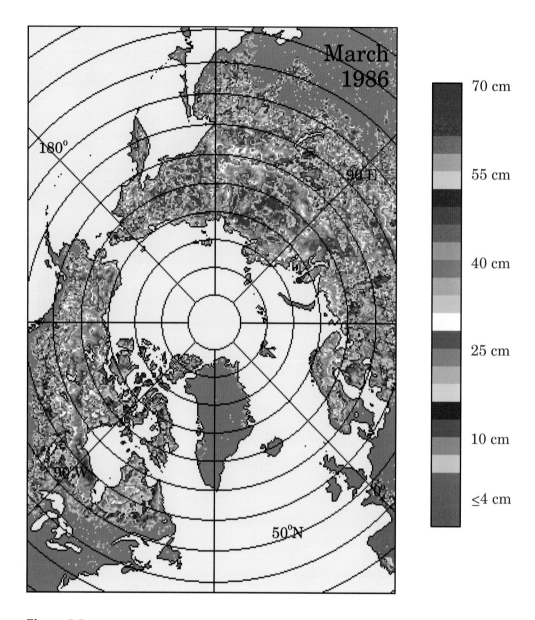

Figure 5.5.
Monthly average snow thicknesses over continental land surfaces in the north polar region for March 1986, calculated from the data of the Nimbus 7 SMMR.

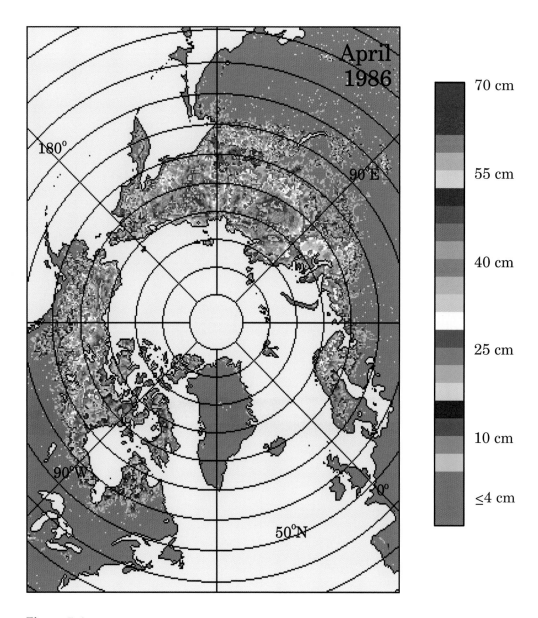

Figure 5.6.
Monthly average snow thicknesses over continental land surfaces in the north polar region for April 1986, calculated from the data of the Nimbus 7 SMMR.

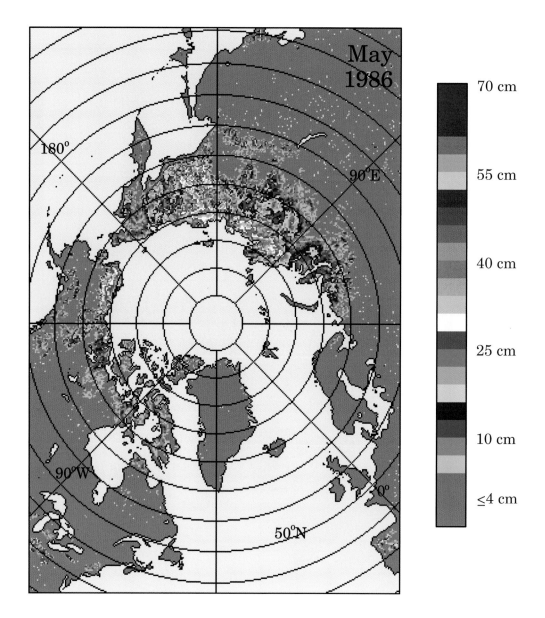

Figure 5.7.
Monthly average snow thicknesses over continental land surfaces in the north polar region for May 1986, calculated from the data of the Nimbus 7 SMMR.

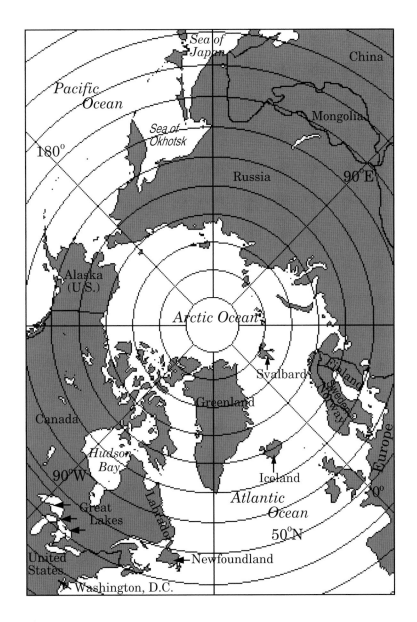

Figure 5.8.
Location map for the north polar region.

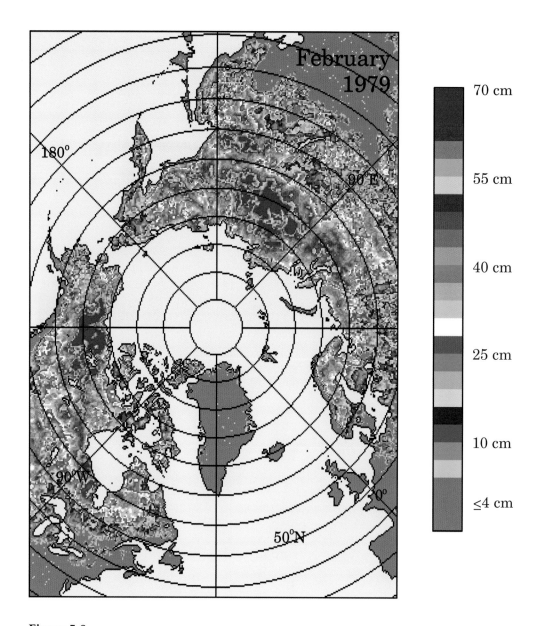

Figure 5.9.
Monthly average snow thicknesses over continental land surfaces in the north polar region for February 1979, calculated from the data of the Nimbus 7 SMMR.

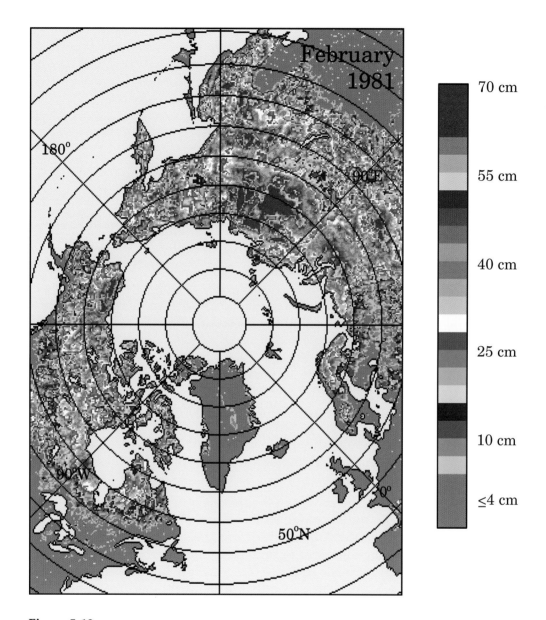

Figure 5.10.
Monthly average snow thicknesses over continental land surfaces in the north polar region for February 1981, calculated from the data of the Nimbus 7 SMMR.

The SMMR radiative data used to obtain the images in Figures 5.4–5.7 and 5.9–5.10 are available through the National Snow and Ice Data Center (NSIDC) in Boulder, Colorado, as are the radiative data from the later SSMI instruments. Together, the SMMR and SSMI data cover the period from late October 1978 until the present.

Questions Regarding the Satellite Imagery

1. In the color scale of Figures 5.4–5.7 and 5.9–5.10, what snow thickness value is associated with each of the following color breaks:
   a. The break between yellow and green.
   b. The break between green and blue.
   c. The break between brown and pink.

2. According to the image in Figure 5.4:
   a. In which one of the following regions did the average February 1986 snow cover show the thickest values: Alaska, Canada, Norway, Finland, or north-central Russia?
   b. In the northern portion of Alaska, what were the approximate average snow thicknesses for February 1986?
   c. What was the approximate average February 1986 snow thickness at 65°N, 110°E (in Russia)?
   d. Along which coast of Hudson Bay were snow thicknesses in February 1986 generally higher: the east coast, the southwest coast, or the northwest coast? Were the thicknesses along that coast generally higher or lower than the snow thicknesses in central Alaska?

3. Using the images in Figures 5.4 and 5.5:
   a. In the section of eastern Russia lying between the Sea of Okhotsk and the Arctic Ocean, was the snow cover in March 1986 generally thicker or thinner than that in February 1986?
   b. In Canada to the east of Hudson Bay, was the snow cover in March 1986 generally thicker or thinner than that in February 1986?
   c. In western Russia north of 55°N and between longitudes 45°E and 60°E, was the snow cover in March 1986 generally thicker or thinner than that in February 1986?

4. Of the months February, March, April, and May 1986, which two have the most similar Northern Hemisphere snow covers according to Figures 5.4–5.7?

5. According to Figures 5.4–5.7, as the Northern Hemisphere snow cover retreated during the spring of 1986:
   a. On a monthly average basis, did the greatest retreat in areal coverage of snow in Russia occur from March to April or from April to May?

    b. On a monthly average basis, did the greatest retreat in areal coverage of snow in Alaska occur from March to April or from April to May?

6. According to Figures 5.9 and 5.10, which of the two years 1979 and 1981 had more extensive February snow coverage in the following regions:
    a. Immediately west of the Great Lakes.
    b. Norway and Sweden.
    c. Finland.
    d. Within Mongolia and southern Russia in the latitude and longitude ranges 45°N–55°N and 90°E–110°E.

7. According to Figures 5.9 and 5.10, which of the two years 1979 and 1981 had thicker February snow coverage in the following regions:
    a. Far eastern Russia, 150°E–180°E.
    b. Northern Norway, Sweden, and Finland.
    c. Northeastern Alaska and northwestern Canada, in the vicinity of 135°W.
    d. Immediately west of the Sea of Japan.

8. Comparing Figures 5.4, 5.9, and 5.10, in each of the following cases was February 1986 more similar to February 1979 or to February 1981?
    a. Snow coverage immediately west of the Great Lakes.
    b. Snow thicknesses in northern Norway, Sweden, and Finland.
    c. Snow coverage in western Canada.

9. Of the three years 1979, 1981, and 1986, which had the most substantial February snow coverage in the region of far eastern Russia extending from the north coast of the Sea of Okhotsk to the Arctic Ocean?

10. Of the three years 1979, 1981, and 1986, which had the most substantial February snow coverage in south central Canada (south of 55°N, west of 90°W)?

CHAPTER **6**

Sea Surface Temperatures and the El Niño

Introduction

Oceans cover approximately 70% of the Earth's surface area and interact with the atmosphere in many ways, as part of the Earth's highly coupled climate system. Over the centuries, the oceans have repeatedly proven to be major obstacles to human exploration and travel, but such problems pale to insignificance when viewed in the context of the larger, positive role of the oceans in the Earth system. The oceans contain the overwhelming majority of the water vital to all of Earth's life forms; they provide the life habitat for hundreds of thousands of species; they transport immense quantities of heat from low to high latitudes; and they serve as a stabilizing factor for climates around the world.

Ocean–atmosphere interactions crucial for the Earth's climate include the exchange of momentum (calculated as mass times velocity), heat, and a variety of chemical elements and compounds (Figure 6.1). Transfer of momentum takes place every time winds generate waves or other water motions (Figure 6.1a). Heat transfers occur wherever a strong ocean–atmosphere temperature difference exists, with the heat going from the warmer to the cooler medium (Figure 6.1b). The rates of transfer depend on the air–water temperature difference and on the surface wind speed; the greatest rates occur with high temperature difference and high wind speed. In view of the high variability of temperatures and winds, the heat transfers between the ocean and atmosphere also vary strongly with location, date, and time of day. For instance, in winter in the Arctic, the heat transfer is predominantly from the water upward to the atmosphere, as the water temperatures are near the freezing point but the winter air temperatures are typically well below freezing.

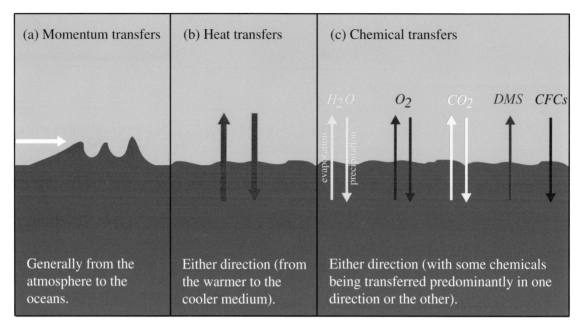

Figure 6.1.
Major ocean–atmosphere exchanges critical to the Earth's climate system.
(a) Momentum transfers, generally going from the atmosphere to the oceans,
as winds generate waves. (b) Heat transfers, proceeding from the warmer to
the cooler medium and hence dependent on location and time. (c) Chemical
transfers, including water vapor (H_2O), oxygen (O_2), and carbon dioxide (CO_2),
each of which has major transfers in each direction, plus dimethylsulfide
(CH_3SCH_3, commonly abbreviated DMS), which is transferred predominantly
from the ocean to the atmosphere, and chlorofluorocarbons (CFCs), which are
transferred predominantly from the atmosphere to the ocean.

Primary amongst the elements and compounds transferred between
the ocean and the atmosphere (Figure 6.1c) is water, with important sec-
ondary transfers including those of oxygen, carbon dioxide, dimethylsulfide, and chlorofluorocarbons. Water is transferred from the ocean to the
atmosphere during evaporation and from the atmosphere to the ocean
during precipitation, evaporation and precipitation both being critical
processes in the climate system. Dimethylsulfide is generally transferred
only in one direction, from the ocean to the atmosphere, after production
by algae in the upper ocean layers. Chlorofluorocarbons are generally
transferred in the opposite direction, from the atmosphere to the ocean,
after having been created by humans and discharged to the atmosphere.
Oxygen is transferred in both directions, although the greater flux on average is upward, from the ocean to the atmosphere. Transfers of carbon
dioxide also occur in both directions. The carbon dioxide transfers have

received particular attention recently because of what is known as the "missing carbon" (or "missing carbon dioxide") dilemma.

The missing carbon dilemma concerns the mismatch between estimates of the average rate at which carbon dioxide is entering the atmosphere (I) from all sources, the average rate at which carbon dioxide is leaving the atmosphere (L), by going into the ocean and elsewhere (in particular, land vegetation), and the average rate at which atmospheric carbon dioxide is increasing (R). The relationship between these three rates should be R = I − L, but estimates of the three values have R considerably lower than I − L, thereby creating a dilemma as to which of the estimates is in error. Conceivably the oceans are taking up more or releasing less carbon dioxide than current estimates suggest.

Ocean surface temperatures (or, in the standard terminology, "sea surface temperatures") tend to be highest in the equatorial regions, where solar heating is greatest on an annual basis, and lowest in the polar regions, where solar heating is least. Ocean currents, however, move large quantities of heat, causing noticeable deviations from a precise correlation between temperature and latitude. For instance, the warm Gulf Stream waters move from low latitudes of the western North Atlantic northeast across the North Atlantic toward Europe (Figure 6.2a) and, along with the North Atlantic Drift extension of the Gulf Stream, result in much higher sea surface temperatures to the west of Great Britain and Scandinavia than would otherwise exist. These higher ocean temperatures in turn yield greater emissions of radiation from the ocean and greater heat transfers to the atmosphere. A result is higher atmospheric temperatures and a much milder northern European climate than would exist without the Gulf Stream influence.

The locations of the Gulf Stream and other major ocean currents are mapped in Figure 6.2a. Just as the Gulf Stream and North Atlantic Drift bring warm waters northeast across the North Atlantic, warming northern Europe, the Brazil Current brings warm waters southward along the western South Atlantic, warming the southern east coast of South America, and the Kuro Current brings warm waters northward along eastern Asia, warming Japan. Conversely, the cold Humboldt and Benguela currents cool the west coasts of southern South America and southern Africa, respectively, and the cool East Greenland and Labrador currents cool the east coasts of southern Greenland and northeastern Canada, respectively. Similar statements could be made about each of the other currents in Figure 6.2a that contain strong north–south components to their flow directions.

Figure 6.2a is necessarily generalized, as currents do not remain constant over time; many show distinct seasonal contrasts as well as interannual variations. In fact, in some regions the currents are omitted in Figure 6.2a explicitly because their seasonal contrasts are so great that the

(a)

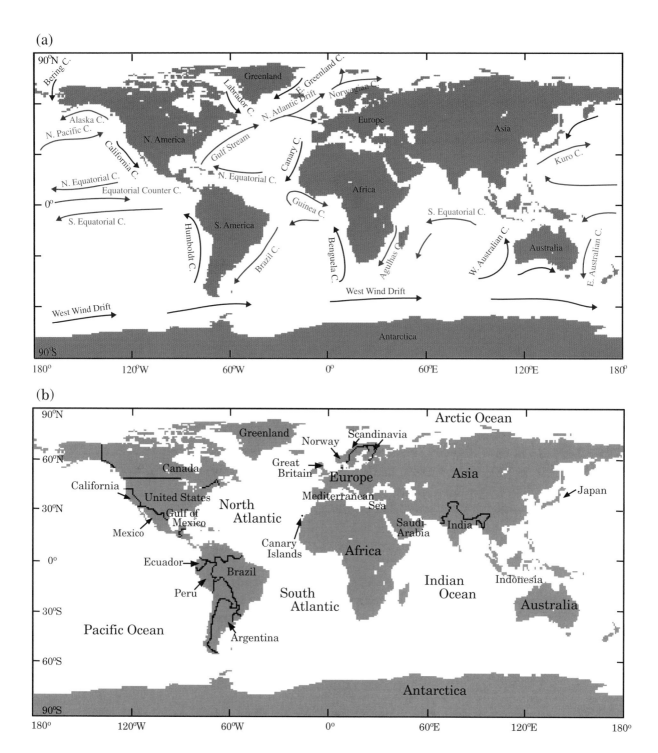

(b)

Figure 6.2.

Location maps. (a) Major ocean currents. "Warm" currents are represented in red, "cold" currents in blue, where the terms *warm* and *cold* refer to the feel of the current in its specific location rather than to precise temperature ranges. For example, the Canary Current is labeled a cold current because it transports waters to the Canary Island region off the west coast of Africa that are colder than the waters in this region in the absence of the current. Its waters, however, are considerably warmer than those of the Norwegian Current, which is labeled a warm current because it transports waters to the west of Norway from much warmer, lower-latitude portions of the North Atlantic. (b) Other locations mentioned in the text and questions.

80

direction of flow is actually reversed from one season to another. This is true, for instance, along the southeast coast of Saudi Arabia, where, in connection with the seasonal wind contrasts often referred to as the Indian monsoon, the current flows to the southwest in winter but to the northeast in summer.

Sea surface temperatures, with their marked responses to solar heating and ocean currents, have a clear annual cycle, clear geographic differences, and some prominent interannual variations, as is also the case for ozone (Chapter 3), sea ice (Chapter 4), snow cover (Chapter 5), and most other climate variables. Overall, sea surface temperatures are higher in summer than in winter and higher at low latitudes than at high latitudes, in spite of the important adjustments produced by ocean currents. One particular sequence of interannual variations has received considerable research attention in recent decades because of its large-scale and somewhat systematic and cyclic nature, along with its apparent relevance to wide-ranging weather conditions. This is the sequence of sea surface temperature changes in the equatorial Pacific associated with the phenomenon termed *El Niño*.

El Niño originally referred to an annual warming of the waters along the coast of Peru. The name *El Niño*, Spanish for *the Christ Child* (or, when left uncapitalized, simply *the boy* or *the child*), arose because the warming generally occurs shortly after Christmas. Usage subsequently became more specialized, however, and the term is now usually restricted to those episodes when the warming is much greater than average and extends well westward into the central equatorial Pacific. The episodes can last through several seasons of the year and do not necessarily begin during the Christmas season.

El Niño is now recognized to be intricately linked to large-scale interannual variations in the atmosphere known as the Southern Oscillation. As a result, the two phenomena are often spoken of together and termed the El Niño/Southern Oscillation, or ENSO. Basically, during non–El Niño conditions, the atmosphere above the equatorial Pacific has strong easterly trade winds (blowing from east to west), and these winds produce water flows that transport masses of warm surface waters to the west. This transport raises sea levels in the west and, to replace the water that has moved westward, encourages upwelling of cold water in the east, along the South American coast. Heavy rains occur in the western Pacific, above the warm surface waters (Figure 6.3a). In contrast, during an El Niño episode, the easterly trade winds are weaker. Consequently, the warm surface waters that have accumulated in the west during non-El Niño conditions no longer receive strong westward forcing from the winds and hence flow (or "slosh" back) eastward; the upwelling of cold waters along the South American coast is slowed or stopped; and heavy rains fall farther east in the central Pacific (Figure 6.3b).

(a) Non-El Niño equatorial conditions

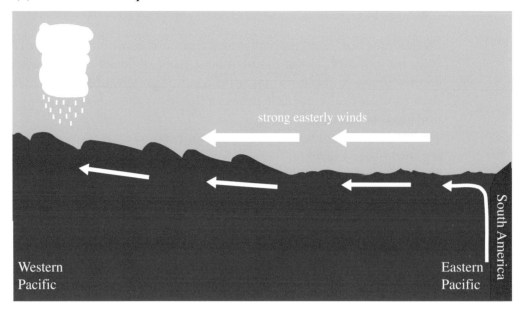

(b) El Niño equatorial conditions

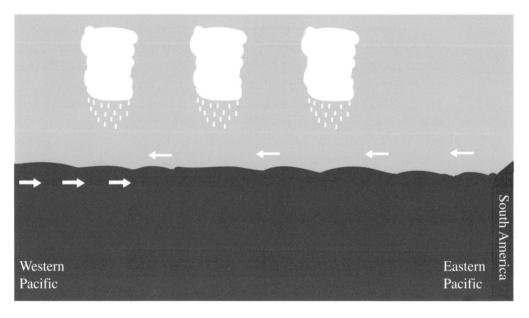

Figure 6.3.
Schematic vertical slices along the equatorial Pacific for (a) Non–El Niño conditions and (b) El Niño conditions. During non–El Niño conditions, strong easterly winds force surface waters westward, leading to a build-up of warm waters in the west and an upwelling of cold waters in the east, along the west coast of South America. During El Niño conditions, the easterly winds are much more relaxed (and perhaps even reversed), the warm waters that had accumulated in the west flow back toward the east, and the upwelling along the west coast of South America is eliminated or at least greatly reduced. The region of greatest atmospheric convection, cloud formation, and resulting precipitation basically follows the movement of the warm waters.

82

El Niños have long been known to have societal consequences to the people of Peru. During El Niño episodes, Peru often experiences heavy rains that can result in unusually abundant growth of land vegetation, a favorable consequence that prior to the development of the major Peruvian fishery industries contributed to the El Niño years being known as "years of abundance." The perceptions of the El Niño are quite different now, however, as the El Niño conditions can severely damage the local fishing industries, by reducing or even stopping the nutrient-laden upwelling along the Peruvian coast. The upwelling brings to the surface the nutrients essential to the fish and other life in the coastal waters.

Over the past few decades, research has linked the El Niño phenomenon to weather conditions across a much vaster region than Peru and its vicinity. In fact, El Niño linkages could extend worldwide and could involve a multitude of events with societal impacts. Among these are droughts in Australia and Indonesia, extreme episodes of the Indian monsoon, hurricanes in the Atlantic, and torrential rains not just in Peru and Ecuador but at least as far north as California. Satellite data have helped clarify the El Niño sequence, with sea surface temperature data in particular helping to reveal the sequence of temperature effects across the Pacific. Successful forecasts of El Niños could help lessen financial impacts, for instance by encouraging farmers to plant crops more appropriate for the precipitation levels of El Niño years.

Review Questions

1. Approximately what percentage of the Earth's surface area is covered by oceans?

2. For each of the following statements, select the appropriate word or group of words in the parentheses that would make the statement correct:
   a. The oceans contain the majority of the Earth's (radiation?; water?), vital to all of Earth's life forms.
   b. The oceans provide the habitat for (hundreds of thousands?; billions?) of species.
   c. On average, the oceans transport heat from (low to high?; high to low?) latitudes.
   d. The oceans serve as a (stabilizing?; destabilizing?) factor for climates around the world.

3. Which quantity is being transferred from the atmosphere to the ocean every time winds generate waves—heat, mass, or momentum?

4. Why are ocean–atmosphere heat transfers in the Arctic in winter predominantly from the water to the atmosphere instead of from the atmosphere to the water?

5. In what major process does water get transferred from the ocean to the atmosphere?

6. Of dimethylsulfide, chlorofluorocarbons, water, and oxygen, for which one are the ocean–atmosphere transfers predominantly from the atmosphere to the ocean?

7. Describe the "missing carbon dioxide" dilemma.

8. What major current is a large part of the reason that the climate of Great Britain is considerably milder than the climates of many regions at the same latitudes?

9. **a.** If changes in ocean circulation were to stop or reverse the Brazil Current (Figure 6.2a), would the climate of coastal Argentina from about 35°S to about 47°S probably become generally warmer or generally cooler than it is now?
   b. Why?

10. **a.** Using the map in Figure 6.2a, at latitudes of 10°S–40°S and 10°N–40°N, are the west coasts of continents generally warmed or generally cooled by the ocean's major currents?
    b. For the same latitude ranges, are the east coasts of continents generally warmed or generally cooled by the ocean's major currents?

11. **a.** What does the acronym "ENSO" stand for?
    b. Of the two parts "EN" and "SO", which one is predominantly associated with the oceans?
    c. Of the two parts "EN" and "SO", which one is predominantly associated with the atmosphere?

12. Comparing El Niño and non–El Niño episodes:
    a. Which have the stronger easterly trade winds?
    b. Which have the lower sea levels in the western Pacific?
    c. Which have the greatest cold-water upwelling along the west coast of South America?

13. **a.** El Niño years used to be known as "years of abundance" in Peru. State one reason why.
    b. Why are El Niño episodes now often instead viewed very negatively by many Peruvians?

Satellite Detection of Sea Surface Temperatures

Prior to satellites, measurements of ocean temperatures were very unevenly distributed. Most of the measurements were taken along commercial shipping routes and hence concentrated in the Northern Hemisphere, particularly in the North Atlantic between North America and Europe. With satellites, data coverage can extend globally, with equal weight given to all points in noncoastal surface waters, whether along shipping routes or not.

Several techniques have been used to measure sea surface temperatures from satellites, based on the amount of radiation emitted by the ocean at various infrared and microwave wavelengths. In this chapter, the examples presented use infrared data from the Advanced Very High Resolution Radiometer (AVHRR) on board the NOAA 9 satellite.

The first AVHRR was a four-channel instrument launched on TIROS-N in October 1978. Subsequent four-channel AVHRRs were launched on NOAA 6 and on successive even-numbered satellites in the NOAA series. In June 1981, the first five-channel AVHRR was launched on the NOAA 7 satellite. Subsequent five-channel AVHRRs have been launched on the successive odd-numbered NOAA satellites, providing a continuing record of AVHRR measurements. In particular, the NOAA 9 AVHRR, data from which are presented in the next section, was launched in December 1984. The five channels on the five-channel AVHRRs include one band (0.58–0.68 micrometers) in the visible region of the electromagnetic spectrum, one band (0.725–1.10 micrometers) in the near-infrared region (near the visible wavelengths), two bands (10.3–11.3 micrometers and 11.5–12.5 micrometers on NOAA 7 and NOAA 9; 10.5–11.5 micrometers and 11.5–12.5 micrometers on NOAA 11) in the longer wavelength thermal–infrared region, and one band (3.55–3.93 micrometers) at intermediate wavelengths within the infrared, sometimes also considered as thermal–infrared. These data are used to examine many variables, amongst them being sea surface temperature.

One of the methods used to derive sea surface temperatures from the AVHRR data is the Multichannel Sea Surface Temperature (MCSST) procedure. In this procedure, the data are first checked to determine which data points are contaminated by clouds and aerosols. The data in the contaminated points are then eliminated, because the presence of clouds or high aerosol amounts prevents the accurate determination of sea surface temperatures (SSTs) by the MCSST method. After the contaminated data have been eliminated, the SSTs are calculated for the rest of the data set by an algorithm of the form: $SST = a_0 + a_1T_1 + a_2T_2$, where T_1 is the AVHRR brightness temperature for the wavelength band 3.55–3.93 micrometers, T_2 is the AVHRR brightness temperature for the wavelength band 10.3–11.3 micrometers (or 10.5–11.5 micrometers, depending upon the instrument), and the coefficients a_0, a_1, and a_2 are values preselected to give SST answers that match known sea surface temperatures very closely in case studies with sea surface temperature data available. In other words, the coefficients are determined empirically rather than theoretically, and the underlying assumption is that the coefficients are appropriate in general and not just for the cases used to determine them. Similarly, the SST equation itself ($SST = a_0 + a_1T_1 + a_2T_2$), like the equations for many variables calculated from satellite data, is not theoretically derived as a precise equation for obtaining the particular variable, in this case sea surface temperature, but instead is assumed to be adequate as long as the coefficients are well chosen. Some of the problems with the MCSST algorithm are that errors are introduced when atmospheric humidity is high and when volcanic eruptions result in heavy aerosol concentrations.

A similar but more elaborate procedure than the MCSST algorithm is now being implemented for the five-channel AVHRR data from the NOAA 7 and subsequent odd-numbered NOAA satellites, in an effort to create an improved sea surface temperature data set for the period from 1981 to the present. This joint NOAA/NASA effort, termed the AVHRR Global Area Coverage (GAC) Oceans Pathfinder Project, involves collaborations also with the University of Miami, which has the primary responsibility for developing the SST algorithm, and the University of Rhode Island, where improved cloud corrections are being developed. NOAA has the responsibility of maintaining the satellite instrument and the data flow from it, while the final data products are produced and quality-controlled at NASA's Jet Propulsion Laboratory (JPL) at the California Institute of Technology. This four-institution collaboration is illustrative of the level of cooperation that often exists in generating major satellite data sets. In some cases the collaborations include even more institutions and several different countries.

Some of the key steps involved in the creation of the SST data sets by the AVHRR GAC Oceans Pathfinder Project are: (1) checking the AVHRR radiances for apparently badly flawed data, which are then eliminated from the generated products (but retained in the original data sets for possible future reconsideration); (2) making adjustments for cloud presence, for instance by eliminating cloud-affected data; (3) intercalibrating the different AVHRR instruments; (4) gridding the data to a rectangular grid on a standard map projection, specifically, an *equal-angle* projection, meaning that angles between lines on the map are equal to the corresponding angles on the spherical Earth; and (5) calculating the sea surface temperatures. The Project processed the 1987–1991 data first, completing these five years by the end of September 1995, then proceeded to processing the post-1991 data and the earlier 1981–1986 data.

The basic SST algorithm used by the AVHRR Oceans Pathfinder Project incorporates data from the two thermal-infrared channels of the AVHRR (Channels 4 and 5), along with an estimate of the SSTs from weekly data, plus three sets of five coefficients determined empirically by comparing selected AVHRR values with temperatures measured at the sea surface using instrumented buoys. Which set of coefficients to use is decided in each case by the value of the difference between the Channel 4 and Channel 5 data. When the value of the difference is less than 0.7, one set of coefficients is used; when it is between 0.7 and 1.8, another set is used; and when it is greater than 1.8, the third set is used. There is nothing magical about the dividing points at 0.7 and 1.8; the basic reason for using three sets of coefficients is simply that the results appear to provide more accurate sea surface temperatures than those derived from any single set of coefficients, as was used in the simpler MCSST procedure. Also, all three sets of coefficients differ for the different instruments; for instance, the NOAA 7 AVHRR has a different set of coefficients than the NOAA 9 AVHRR has. Future refinements to the algorithm could include revisions to the 0.7 and 1.8 cutoffs for the coefficient selection and lessening of the abruptness with which the coefficients change at these cutoffs. A more fundamental change in the algorithm could involve a revision in the basic equation, perhaps avoiding the use of multiple sets of coefficients altogether. Changes and updates to the Pathfinder data set can be found through the world wide web at the uniform resource locator (URL) for the SST Pathfinder effort, http://podaac-www.jpl.nasa.gov/sst.

Review Questions

1. In the two centuries prior to satellite observations, why were there more sea surface temperature data for the North Atlantic than for most other major regions of the world's oceans?

2. What types of data are collected by the Advanced Very High Resolution Radiometers (AVHRRs)?

3. What does the "near" refer to in "near-infrared"?

4. Is the Multichannel Sea Surface Temperature (MCSST) procedure, in which SSTs are determined by calculating SST $= a_0 + a_1T_1 + a_2T_2$, an empirical or a strictly theoretical formulation?

5. Name two atmospheric conditions that introduce errors into the calculation of sea surface temperatures using the MCSST algorithm.

6. What is the purpose of the AVHRR Global Area Coverage (GAC) Oceans Pathfinder Project?

7. The AVHRR GAC Oceans Pathfinder Project, like many satellite data projects, involves collaborations amongst at least four major institutions, located distantly from each other.
   a. Name a key advantage of having such wide-ranging collaborations for obtaining data products from satellite observations.
   b. Name a key disadvantage of having such collaborations.

8. What is an "equal-angle" map projection?

9. Considering that the algorithm used by the AVHRR GAC Oceans Pathfinder Project incorporates coefficients determined by comparing selected AVHRR values with temperatures measured directly at the sea surface, is this algorithm partly empirical or strictly theoretical?

Satellite Images of Sea Surface Temperatures

Figures 6.4 and 6.5 present seasonal monthly average sea surface temperatures derived by the AVHRR GAC Oceans Pathfinder Project from the NOAA 9 AVHRR data for 1990. Figure 6.4 shows results for March, at the end of the Northern Hemisphere winter and Southern Hemisphere summer, and for June, at the end of the Northern Hemisphere spring and Southern Hemisphere autumn. Figure 6.5 shows results for September, at the end of the Northern Hemisphere summer and Southern Hemisphere winter, and for December, at the end of the Northern Hemisphere autumn and Southern Hemisphere spring. White on the figures indicates missing data over water, and black indicates land.

Each of the four images in Figures 6.4 and 6.5 reveals a clear overall gradient of temperatures from warm in the equatorial regions to cold in the polar regions. The warmest temperatures indicated are 30°C–35°C, and the coldest are about –2°C, at which temperature the water should be freezing into sea ice. The images also show that temperature values along the same latitude curve tend to be close to each other and that the strongest deviations from such "zonal" behavior occur near continental coasts. For instance, cold water extends northward along the western coasts of southern South America and southern Africa and southward along the east coast of northern North America, all readily understandable, in a qualitative sense, from the flow directions of the Humboldt, Benguela, East Greenland, and Labrador currents (Figure 6.2a). Differences amongst the four months reflect the basic seasonal cycle, with somewhat warmer temperatures in the Northern Hemisphere in June and especially September than in December or March, and somewhat warmer temperatures in the Southern Hemisphere in December and especially March than in June and September (Figures 6.4 and 6.5). Additional spatial patterns and seasonal differences are highlighted in the questions.

To illustrate differences in the patterns of sea surface temperatures between El Niño and non–El Niño episodes, Figure 6.6 presents monthly average May images for 1987 and 1988; the former occurred during an El Niño episode and the latter did not. Overall, sea surface temperatures and temperature patterns around the globe are quite similar in the two Mays, but in the non–El Niño year, 1988, there is a much more prominent streak of relatively cold temperatures extending along the equatorial Pacific westward from the coast of South America and cooler temperatures also in the central equatorial Pacific and in the equatorial Indian Ocean. Seen in a global perspective, as in Figure 6.6, the equatorial streak of cold water in the eastern Pacific in a non–El Niño period (for example, May 1988) can be a prominent feature, narrow, well-defined, and extending a considerable distance across the Pacific. In fact, in May 1988 the feature extended

March 1990

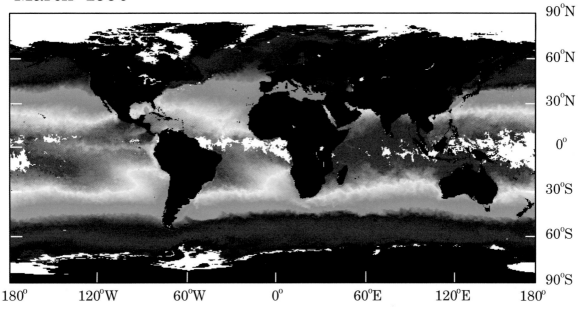

June 1990

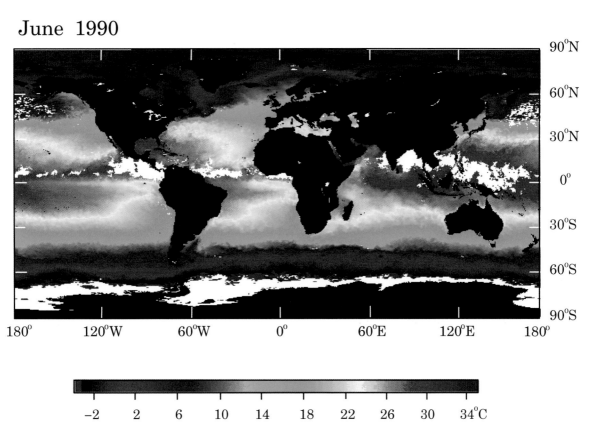

Figure 6.4.
Monthly average global sea surface temperatures for March 1990 and June 1990, calculated from the data of the NOAA 9 AVHRR. The images were obtained digitally from the AVHRR GAC Oceans Pathfinder Project, with tick marks and labels added later.

September 1990

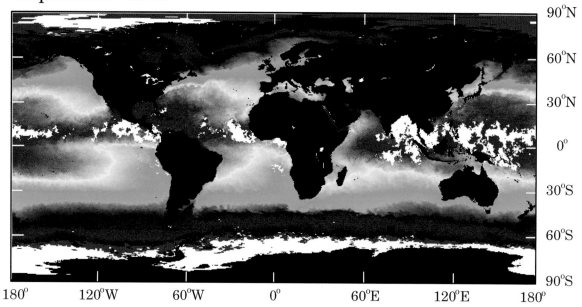

December 1990

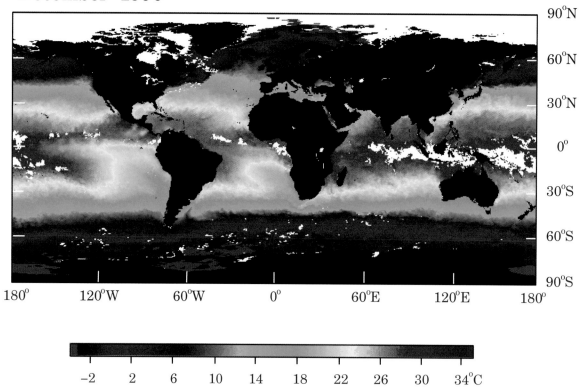

Figure 6.5.
Monthly average global sea surface temperatures for September 1990 and December 1990, calculated from the data of the NOAA 9 AVHRR. The images were obtained digitally from the AVHRR GAC Oceans Pathfinder Project, with tick marks and labels added later.

May 1987

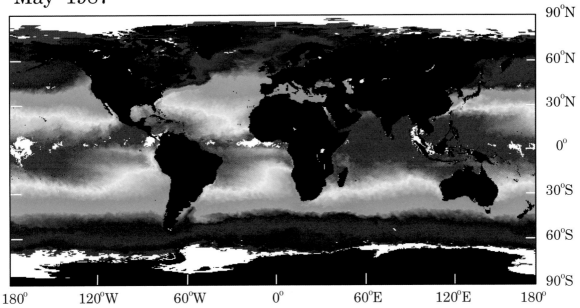

May 1988

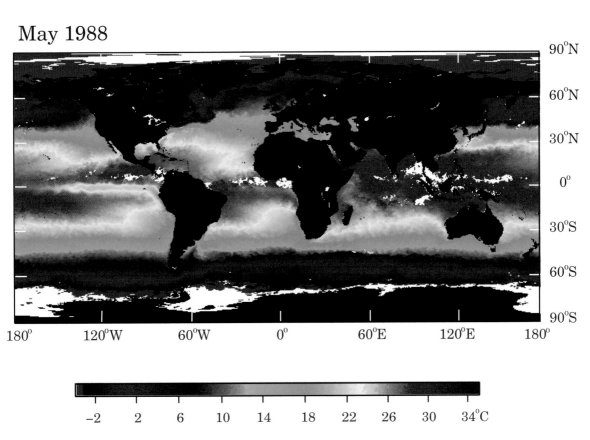

Figure 6.6.
Monthly average global sea surface temperatures for May 1987 and May 1988, calculated from the data of the NOAA 9 AVHRR. The images were obtained digitally from the AVHRR GAC Oceans Pathfinder Project, with tick marks and labels added later.

across a distance far exceeding the distance from the east to the west coast of the United States (Figure 6.6).

The NOAA AVHRR data used to obtain the images in Figures 6.4–6.6 are available through the world wide web at http://podaac-www.jpl.nasa.gov/sst, as part of the NOAA/NASA AVHRR Oceans Pathfinder Project.

Questions Regarding the Satellite Imagery

1. a. According to Figures 6.4 and 6.5, in which of the four months shown (March, June, September, and December 1990) was the sea surface temperature in the eastern Mediterranean Sea the lowest? What was the approximate temperature?

   b. In which of the four months shown was the sea surface temperature in the Mediterranean Sea the highest? What was the approximate average temperature?

   c. What is the approximate average sea surface temperature difference between the warmest and the coldest months in the eastern Mediterranean?

   d. Where you live, are the differences in air temperatures between the warmest and coldest of the four months March, June, September, and December greater than or less than the sea surface temperature differences between the warmest and coldest months in the eastern Mediterranean?

2. a. In which of the four months March, June, September, and December 1990 was the average sea surface temperature in the western Gulf of Mexico the lowest?

   b. In which of the four months was the average sea surface temperature in the Gulf of Mexico the highest?

3. a. According to Figures 6.4 and 6.5, in which of the four months shown were the sea surface temperatures along the southern coast of Australia the lowest? What was the approximate temperature?

   b. In which of the four months shown was the sea surface temperature along the southern coast of Australia the highest? What was the approximate temperature?

   c. What is the approximate sea surface temperature difference between the warmest and the coldest months along the southern coast of Australia?

4. What aspect of the North Atlantic sea surface temperature fields, apparent in each of the images of Figures 6.4 and 6.5, is readily explained by the existence of the Gulf Stream?

5. In which months in Figures 6.4 and 6.5 are the waters along the east coast of southern Africa (0°S–30°S) generally warmer than those along the west coast?

6. The Benguela Current runs along the west coast of southernmost Africa. Without referring back to the map of ocean currents, determine from the images of Figures 6.4 and 6.5 whether the flow is basically south-to-north or north-to-south, and explain how you can tell. (Then check Figure 6.2a to confirm your answer.)

7. The Humboldt Current runs along most of the Southern Hemisphere portion of the west coast of South America. Again, as in Question 6, without referring back to the map of ocean currents, use the images of Figures 6.4 and 6.5 to answer whether the flow is basically south-to-north or north-to-south. (Then confirm your answer with Figure 6.2a.)

8. **a.** In Figure 6.6, which three of the following six locations show the greatest sea surface temperature differences between May 1987 and May 1988: along the east coast of southern South America, along the southern coast of Africa, along the equator in the eastern Pacific, along the equator in the Indian Ocean, in the North Atlantic from Europe west to North America, or in the Gulf of Mexico?

   b. For each of the three selected locations, indicate which of the two years had the higher sea surface temperatures.

9. **a.** In the May 1988 image of Figure 6.6, under non–El Niño conditions, where are the Earth's lowest equatorial sea surface temperatures?

   b. Are these temperatures higher or lower than those for the same month in the Gulf of Mexico?

   c. Are they higher or lower than those off the southeast coast of South America?

CHAPTER 7

Land Vegetation

Introduction

Vegetation covers much of the Earth's land surface and is a major source of sustenance for humans and more than a million other animal species. It is also, for humans, a source of fuel, clothing materials, writing materials, and building materials, the latter category being true for other species as well, including beavers building dams and birds building nests. Land vegetation has been an important part of the Earth's biological web for more than 400 million years and comes in a wide variety of forms, from majestic redwood forests to the smallest of plants, invisible to the naked eye.

Major Habitat Types

The tremendous variety of land vegetation throughout the world can be broadly categorized into five major habitat types: (1) forests, where trees are sufficiently close that the ground is largely shaded from the direct rays of the Sun, (2) savannas, where trees are spaced singly or in small groups, interspersed with grasses, shrubs, or other low-lying plant life, (3) grasslands, consisting largely of herbs, including grasses and grasslike plants, (4) deserts, where plants (including herbs and lichens, the latter consisting of algae and fungi living together in a single plant structure) are thinly dispersed over a largely unvegetated surface area, and (5) tundra, consisting of a mixture of grasses, flowering herbs, lichens, mosses, sedges, and sometimes low-lying shrubs, all growing in a cold environment where the ground is likely to be frozen year-round beneath a depth of about half a meter. Tundra can occur at high elevations irrespective of latitude and at all elevations in high latitudes. In either case, the growing season is too short to support trees. Figure 7.1a shows the geographic distributions of tundra, deserts, and two major forest types; Figure 7.1b provides a map of the locations of places named in the text.

(a)

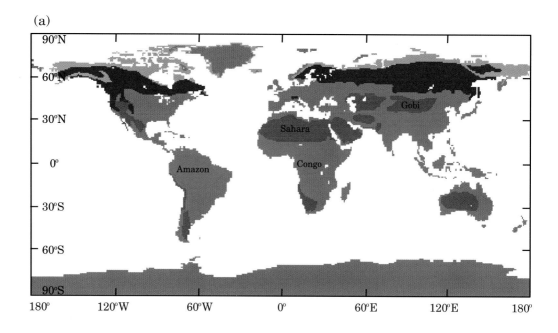

(b)

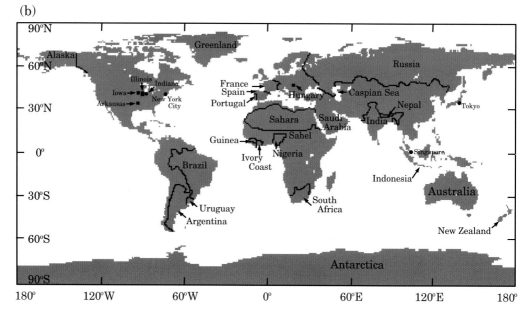

Figure 7.1.
Location maps. (a) Approximate geographic distributions of the Earth's major deserts (in brown), tundra (in pink), needleleaf forests (in blue), and rainforests (in green). Labels are provided for two major named deserts, the Gobi and the Sahara, and two major named rainforests, those of the Congo and Amazon lowlands. (b) Other locations mentioned in the text and questions.

Each of the major habitat types contains tremendous variety and could be subdivided in many ways. In the next two paragraphs, these possibilities are illustrated by listing a few subdivisions for the forest and grassland habitats. First, however, this paragraph presents some terminology to clarify the forest subdivisions. Specifically, trees can be classified in terms of leaf structure as *broadleaf* or *needleleaf* (Figure 7.2) and in terms of whether or not they retain their leaves throughout the year as *evergreen* or *deciduous* (Table 7.1). Deciduous trees drop their leaves either for the cold season or for the dry season, depending upon the location and corresponding climatic conditions for the particular tree. Broad leaves lose con-

Table 7.1

Sample trees, categorized as deciduous or evergreen and as broadleaf or needleleaf

	Broadleaf		Needleleaf	
Deciduous	Apple	Ebony	Bald cypress	
	Ash	Elm	Dawn redwood	
	Baobab	Hickory	Larch	
	Beech	Maple	Swamp cypress	
	Birch	Pagoda		
	Cannonball	Sycamore		
	Chestnut	Willow		
Evergreen	Banyan	Mahogany	Cedar	Pine
	Cinnamon	Mandarin	Cypress	Redwood
	Cocoa	Mango	Hemlock	Sequoia
	Holly	Papaya	Juniper	Spruce
	Laurel	Rubber	Fir	Yew
	Lemon	Tea		
	Lime	Teak		

Figure 7.2.
Samples of the two major categories of leaves: (a) broadleaf, (b) snow-covered
needleleaf. [Photographs from the author.]

siderable moisture to the atmosphere, so their removal provides a mechanism for the tree to reduce moisture losses. Deciduous trees are largely dormant during the period without leaves. The brilliant coloration of many of these trees shortly before the leaves fall results as various chemical substances withdraw from the leaves into the rest of the tree for winter storage (Figure 7.3). Evergreen trees can also shed their leaves, although they do not lose their leaves on a whole-tree, annual basis as deciduous trees do; as a result, evergreen trees can remain green throughout all seasons.

Amongst the subdivisions of the forest habitat type are equatorial rainforests, tropical rainforests, temperate rainforests, monsoon forests, temperate deciduous forests, broadleaf evergreen forests, and needleleaf (or coniferous) forests, each with its own climatic requirements and resulting geographic distribution. The equatorial rainforests, located in the Amazon lowland of South America, the Congo lowland of Africa, the African coastal zone from Nigeria to Guinea, and Indonesia and surrounding coastal areas (Figure 7.1a), are subject to continuous warmth and abundant precipitation. Such tree-favorable conditions result in a huge variety of tree types; in fact, at some locations in the equatorial rainforests, 3000 species of trees coexist in a single square mile. In sharp contrast, in the North American and Eurasian needleleaf forests, where conditions are far colder and drier, one or two species can extend for vast expanses without any mixture of other tree types. The needleleaf trees accommodate themselves to the harsher climatic conditions partly through their narrow needles, which lose much less moisture to the atmosphere than broad leaves under the same conditions. Among the world's needleleaf forests are the boreal forests spreading across North America and Eurasia from 45°N to 70°N, consisting of evergreen conifers such as spruce, fir, and pine in North America, Europe, and western Asia and deciduous larch in central and eastern Asia. Further equatorward, the needleleaf evergreen forests of west coastal North America include Douglas firs and redwoods; the latter are among the tallest trees in the world, some attaining heights exceeding 100 meters (325 feet), with correspondingly large circumferences of about 20 meters. The mechanism of leaf-dropping for moisture retention allows deciduous broadleaf trees to extend further poleward than do many types of evergreen broadleaf trees, although generally not as far poleward as the poleward portions of the needleleaf boreal forests.

Two major subdivisions of grasslands are tall-grass prairies and short-grass steppes. The grasslands occur primarily in mid-latitude and subtropical climates with little or no water surplus, hence accounting for the general absence of trees except along water courses, where groundwater can often support tree growth. Prior to the nineteenth century, grasslands spread over vast expanses of North America, supporting herds of ante-

Figure 7.3.
(a) Green leaves from a broadleaf tree beginning to change color as the nutrients retreat from the leaves and withdraw into the tree core. (b) A deciduous broadleaf tree with its colors changed and some of its leaves fallen to the ground. [Photographs from the author.]

100

lope and bison and providing a habitat for the American Plains Indians. In some of the more humid sections, including the regions that are now Iowa, Indiana, and Illinois, the grasses often grew as tall as 3 meters (10 feet). Accounts written by inhabitants of and sojourners in these *tall-grass prairies* told of impressive vistas when, under conditions of a gentle breeze, seemingly endless expanses of tall grasses would flow in unison. Similar sights were seen in Eurasia from Hungary to northern China, in South America in the Pampas of Uruguay and Argentina, and in Africa in the Veldt of South Africa. In the drier grassland sections of continental interiors, heights of the grasses are more typically about 15–30 centimeters (6–12 inches), and the corresponding term for the habitat is *short-grass steppe*.

Human Impacts on Vegetation

Humans have affected vegetation in some way ever since our species first appeared. Initially the effects were minor, as the small human population joined other animal species in trampling vegetation underfoot and eating from vegetal growth. About 10,000 years ago, however, a fundamental change occurred as humans developed agriculture and began intentionally cultivating individual types of vegetation. At the same time, hunters also affected the landscape, for instance through their use of fire to drive and concentrate game animals. Much more recently, as the human population has soared, human impacts on the Earth's vegetation have soared as well. Because of the typically organically rich soil of the mid-latitude grassland regions, humans have taken great advantage of them, in the process almost completely transforming the tall-grass prairies and transforming to a lesser extent the drier short-grass steppes, where irrigation is more often needed for successful agriculture. The wild prairie grasses have been replaced by domesticated wheat, corn, and barley, and the regions have become the "breadbaskets" of the world. These almost total transformations of grasslands by humans have received little attention from environmental groups. After all, the grassland transformations are largely completed and have resulted in hugely favorable consequences for feeding the human population. In contrast, the widescale destruction of forested areas, which is still ongoing and which is seen to have very negative consequences, has received a great deal of attention.

Forests have been destroyed by humans predominantly for two purposes: to obtain timber for a variety of uses, including fuel, paper, and building materials, and to clear land for agriculture, settlements, and other human constructs, including roads and buildings. Significant portions of the prior regional forest cover were destroyed, through human activities, in Europe by the seventeenth century, in North America in the seventeenth through nineteenth centuries, and in the tropical rainforests of South America, Africa, and Indonesia, ongoing in the second half of the twentieth century. Also ongoing in recent decades has been the further decimation of some mid- and high-latitude forests, including the largest national forest in the United States, Alaska's Tongass forest; but the tropical rainforests have received particular attention because of the scale of their destruction and the diversity of species living within them. By the end of the 1980s, global tropical deforestation was proceeding at a rate of about 1.8% per year, with approximately 142,000 square kilometers of rainforest eliminated annually. This area is slightly larger than the areas of either the state of Arkansas or the country of Nepal. In some countries, such as Nigeria and the Ivory Coast, the deforestation rate has exceeded 14% per year. Altogether, over the last 300 years, an estimated 15%–20% of the world's forest area, amounting to about 8 million square kilometers, has been destroyed by human activities. Over the past 10,000 years, perhaps one-third of the Earth's forested area that existed 10,000 years ago has been eliminated. Purposeful replanting of trees (along with haphazard regrowth over abandoned fields and settlements) has helped reverse the trend in portions of the United States and elsewhere, showing that such a reversal is possible; but the continuing destruction of the tropical rainforests in particular remains a serious problem.

Although tropical rainforests cover only a small percentage (about 6%, or 8–9 million square kilometers) of the Earth's land area, they house a tremendous diversity of life, as illustrated previously with the mention of 3000 tree species growing in a single square mile. Noted biologist Edward O. Wilson estimated in 1992 that more than half the species of organisms on Earth live in the rainforests. Furthermore, thousands of those species are believed to live nowhere else. In addition to their importance in the overall scheme of life on Earth, these species-diverse rainforests provide many immediate practical benefits, among the most important of which is as a source of the core substances for many known medicines and probably several times as many potential but as yet undiscovered medicines. Extinction of the species that contain these substances would damage the practice of medicine as well as the rainforests and the global ecosystem. Although forests themselves can often be reestablished, species that become extinct as a result of deforestation will almost certainly remain extinct (notwithstanding the incredible possibilities showcased in such science fiction tales as the book and movie *Jurassic Park*, in which long-extinct dinosaur species are brought back to life).

Deforestation also has important effects on climate, for instance through its tendency to increase the carbon dioxide content of the atmosphere. Young, growing forests take in carbon dioxide and release oxygen during the critical process of photosynthesis, during which radiation acts on carbon dioxide (CO_2) and water (H_2O) to produce the food of carbohydrates ($C_6H_{12}O_6$) and the waste product of oxygen (O_2) (Figure 7.4a).

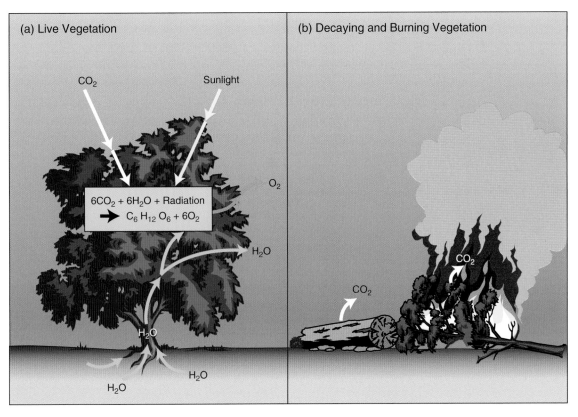

Figure 7.4.
Critical carbon dioxide (CO_2) and water (H_2O) flows to and from live and decaying vegetation. (a) Absorption of carbon dioxide and transmission and absorption of water by live vegetation. The chemical equation presented is the equation for photosynthesis. (b) Release of carbon dioxide to the atmosphere by decaying and burning vegetation.

Live vegetation thus helps reduce levels of carbon dioxide in the atmosphere. In contrast, burning and decay of vegetation both release carbon dioxide to the atmosphere (Figure 7.4b), contributing to the rise in atmospheric carbon dioxide and thereby to the possibility of widespread warming. The net loss of tropical rainforests from 1850 to 1980 is estimated to have resulted in the addition of 90–120 billion metric tons (90–120 trillion kilograms) of carbon dioxide to the atmosphere. This amount compares with an estimated 165 billion metric tons of carbon dioxide added to the atmosphere through the burning of coal, oil, and gas.

Because the estimates of carbon dioxide releases to the atmosphere resulting from deforestation are large and contain considerable uncertainty, increased understanding of deforestation could shed light on the

issue of the missing carbon dilemma mentioned in Chapter 6. This dilemma centers on the fact that the carbon content of the atmosphere has
risen less than estimates of the rates of input to and output from the atmosphere would suggest it should have risen. The magnitude of the dis-
crepancy would be affected by any major adjustments to the estimated
inputs, for instance from industrial sources or decaying vegetation, or to
the estimated outputs, for instance to the ocean or to photosynthesis in
live land vegetation.

Deforestation and other changes in land cover also affect climate
through changes in the albedo (reflectivity) of the surface, changes in the
amount of evapotranspiration (transmission of water to the atmosphere
from plants), and changes in soil erosion, which tends to be increased as
vegetation is removed (because of the role plant roots play in holding the
soil in place). Regarding evapotranspiration, rooted plants take water
from the soil, use a portion of it in photosynthesis, and release a portion
of it to the atmosphere as water vapor (Figure 7.4a). When plants are removed, through deforestation or otherwise, this mechanism of water flow
to the atmosphere is cut off, tending to reduce atmospheric moisture.
Such drying of the atmosphere could result in reduced rainfall in the surrounding regions as well as in the deforested areas, and this could have
the additional consequences of limiting plant growth and agricultural productivity. The drying, along with the crucial fact that rainforest soils (in
marked contrast to the soils of the mid-latitude grasslands) often are inappropriate for agriculture, contributes to the abandonment of many
farms within a few years after tropical forest areas have been cleared for
them. As long as such farming is done only on a small scale, as was generally the case in the past, this sequence can work successfully. Basically,
when people cut down and burn the vegetation in a small area, the soil
enrichment from the burned biomass can allow a few seasons of successful crop harvesting, after which the soil nutrients are depleted, the farmers abandon the field and move on to create a new plot, and the surrounding rainforest reinvades the abandoned field, with no permanent loss of
rainforest area. A central difficulty today, however, is that the deforestation is no longer small scale. With modern equipment and the construc-
tion of major highways and innumerable smaller roads through the forests, the areas cleared are often too large to allow easy reestablishment of
the rainforest even after the fields are abandoned.

Other notable vegetation-related changes that have occurred within
the time frame of the satellite record include the expansion and contraction of major deserts. For instance, the largest of the Earth's dry deserts,
the 7–10-million square-kilometer Sahara Desert of northern Africa (Figure 7.1a), expanded southward over the period 1980 to 1984, then con-
tracted back somewhat in subsequent years. The expansion in the 1980–

1984 period and the preceding decade of the 1970s has been attributed partly to drought and partly to land mismanagement along the edges of the desert, particularly depletion of the groundwater, overgrazing, excess cultivation, and cutting of firewood. To the south, the desert conditions of the Sahara make a gradual transition to the semiarid steppe conditions of the Sahel. Although neither *desert* nor *steppe* has a widely accepted precise definition, both habitats are considerably drier than most forested areas, and deserts are drier than steppes. Rainfall in much of the Sahara is typically less than 10 centimeters per year, whereas in the Sahel it typically falls within the range of 20–40 centimeters per year. At these low levels of precipitation, the amount of rainfall is a primary determinant of vegetation growth. In contrast, average annual rainfall is about 102 centimeters in New York City, about 175 centimeters in Tokyo, and about 238 centimeters in Singapore.

Because of the importance of vegetation to the exchanges of water, energy, and carbon dioxide between the land surface (including its vegetation cover) and the atmosphere, considerable effort has been spent in the 1980s and 1990s on incorporating vegetation into general circulation models of the global climate. Including vegetation appropriately can improve the simulations of present climate and of future climate change.

Review Questions

1. What are some of the important ways in which humans use vegetation?

2. In addition to forests, what are four other major habitat types for land vegetation?

3. Consider a deciduous broadleaf tree. Of what value is it to the tree to have its leaves fall every year?

4. How does the leaf structure of a needleleaf tree help it to accommodate to harsh climatic conditions?

5. Order the following habitat types in terms of how humid the climate is likely to be, starting with the most humid: prairies, steppes, rainforests, deserts.

6. What fundamental change in human impacts on vegetation occurred about 10,000 years ago?

7. In the decades before the regions now known as the "breadbaskets of the world" acquired that label, what was their most widespread basic habitat type: forest, savanna, grassland, desert, or tundra?

8. In the past several hundred years, as humans have cut down trees in many forested areas around the world, what have been the two major broadscale reasons for this deforestation?

9. **a.** E. O. Wilson has estimated that as of 1989, tropical rainforests covered an area of 8 million square kilometers and were being reduced at a rate of approximately 142,000 square kilometers per year. Assuming these numbers are correct, how long could this deforestation rate continue, starting in 1989, before all the remaining rainforests would have disappeared?
   b. In which decade would this be?

10. Does purposeful replanting of trees help **reverse the trend** toward further decreases in total forest area?

11. How have the rainforests helped the practice of medicine?

12. After deforestation has occurred, is it easier to recreate a forest in the same location or to **recreate the** species that became extinct because of the deforestation?

13. How does photosynthesis **affect the** carbon dioxide content of the atmosphere?

14. Has deforestation over the past 150 years contributed significantly to the increase of carbon dioxide in the atmosphere?

15. Many of the global estimates regarding carbon dioxide flows to and from the atmosphere contain considerable uncertainty and might eventually be adjusted significantly as more is learned. For each of the following possible adjustments, indicate whether it would tend to increase the missing carbon dilemma or reduce it:
    a. More carbon dioxide inputs to the atmosphere from decaying vegetation.
    b. Fewer carbon dioxide inputs to the atmosphere from decaying vegetation.
    c. More photosynthesis.
    d. Less photosynthesis.

16. **a.** Does deforestation tend to increase or to decrease soil erosion in the deforested area?
    b. Why?

17. Does deforestation tend to increase or to decrease the amount of moisture in the air in the immediate vicinity?

18. **a.** What is the largest dry desert on Earth today? (*Note:* Antarctica, much of which is a desert in terms of the very low levels of precipitation, is larger. However, Antarctica is considered a "wet desert" rather than a "dry desert," because the surface consists of ice and snow rather than dry soil or sand.)
    b. List some of the human activities that likely contributed to a temporary expansion in the size of this desert in the 1970s and early 1980s.

Satellite Detection of Land Vegetation

As with most of the variables highlighted in this book, vegetation has been examined with data from a variety of satellite instruments. Some of the primary instruments used so far to examine vegetation are the Landsat Thematic Mapper and Multispectral Scanner, which obtain detailed information at spatial resolutions of 30 meters and 80 meters, respectively, and the Advanced Very High Resolution Radiometer (AVHRR), which obtains information at the coarser resolution of about 1.1 kilometers. The coarser resolution of the AVHRR data allows large-scale coverage on a much more frequent basis than is possible with Landsat. Hence, the Landsat data are often more valuable for detailed local studies, whereas the AVHRR data are more appropriate (than the Landsat data) for global studies. The passive-microwave Nimbus 7 Scanning Multichannel Microwave Radiometer (SMMR) and DMSP Special Sensor Microwave Imagers (SSMIs) have also been used for global vegetation studies, at the much coarser resolution of about 30 kilometers; but here the samples presented will be from the AVHRR data, which remain the satellite data most widely used for global-scale vegetation mapping.

As mentioned in Chapter 6, where the AVHRR data are used for mapping sea surface temperatures, the AVHRR transmits data in the visible and infrared portions of the electromagnetic spectrum, with a spatial resolution of about 1.1 kilometers directly below the satellite. The NOAA satellites carrying the instrument fly at an altitude of about 850 kilometers and orbit the Earth 14.1 times each 24 hours, obtaining global coverage over that time period.

Vegetation was not one of the variables originally intended for study with the AVHRR, but the instrument was changed in 1979 in order to improve its ability to distinguish clouds from land, and this change additionally opened possibilities for examining vegetation. Specifically, the earlier AVHRR 0.55–0.90-micrometer channel was replaced by a narrower channel spanning the visible wavelengths from 0.58 micrometers to 0.68 micrometers. The modified channel, in conjunction with the AVHRR's 0.725–1.10-micrometer near-infrared channel, has been found to be quite useful for vegetation studies.

A widely used measure of vegetation activity obtained from the AVHRR visible and near-infrared data, and the one used here, is the Normalized Difference Vegetation Index, abbreviated by the acronym NDVI. The NDVI is defined as the ratio $(I - R)/(I + R)$, where I is the measured radiance at the AVHRR's near-infrared wavelength band (0.725–1.10 micrometers) and R is the measured radiance at its "red" wavelength band (0.58–0.68 micrometers, which includes orange as well as red wavelengths). The word *Difference* in the name Normalized Difference Vegetation Index refers to the I – R difference in the numerator of NDVI, and *Normalized* refers to the fact that this difference is divided by the sum I + R. For both I and R, the overwhelming majority of the radiation measured is radiation that has entered the Earth–atmosphere system from the Sun and been reflected back toward space. Healthy green vegetation reflects most of the near-infrared radiation incident on it but very little of the red radiation (Figure 7.5), leading to high values for NDVI. In fact, the NDVI values for vegetation are generally higher than for any other common Earth surfaces, helping to distinguish vegetation in the AVHRR data and thereby explaining why NDVI is useful in vegetation monitoring. Instead of getting reflected, red radiation tends to be absorbed by vegetation, in which it contributes (with other visible radiation) to powering photosynthesis, hence making the absorption of visible radiation critically useful. The lack of absorption of near-infrared radiation by vegetation is also useful, as it helps the vegetation avoid the overheating that would likely result if both visible and near-infrared radiation were strongly absorbed.

Because the value of R tends to decrease as the green vegetation increases (the more green vegetation, the more the red radiation is absorbed rather than reflected) and the value of I tends to increase as the vegetation increases, a high value of NDVI suggests considerable vegetation. Typical NDVI values are about 0.6 for tropical rainforests, 0.05 for deserts, and −0.3 for cloud-free water. However, both floods and droughts generally lower the NDVI values, and when a region is snow-covered, the effect of the snow on R and I makes NDVI no longer appropriate as a vegetation index. Because of these and other limitations of the NDVI measure, various corrections are now being applied or developed to create enhanced NDVI-related products, corrected for such factors as soil background and atmospheric contamination.

An additional AVHRR channel, at thermal-infrared wavelengths of 11.5–12.5 micrometers, is used in the NDVI algorithm to identify cloud-obscured data and screen them out of the final data product. The screening is done by eliminating data from locations and times when the 11.5–12.5 micrometer band has a brightness temperature below 273 K (or, for Africa only, below 285 K). This technique correctly screens out most cloud-contaminated data, although it also eliminates data from some cloud-free areas with particularly cold surface temperatures.

In contrast to the monthly data for ozone, sea ice, snow, and sea surface temperature presented in Chapters 3–6, the monthly NDVI values are not monthly averages but instead are the maximum NDVI values (at each location) for the month. This use of maximum values is because the complicating factors (e.g., clouds) that lead to erroneous NDVI values almost all have the effect of lowering the NDVI, so that the highest NDVI value is more likely than the average value to give a good indication of the vegetation cover.

Figure 7.5.
Schematic diagram illustrating the tendency for healthy green vegetation to reflect most of the near-infrared radiation incident on it but primarily only the green portion of the incident visible radiation; most of the red visible radiation is absorbed instead. The combined reflection of most of the near-infrared radiation but very little of the red radiation and the contrasts in this respect with other surface types are central to the theory behind the mapping of vegetation from the AVHRR satellite data.

The particular NDVI data set used for the illustrations in the following section is contained on a set of five CD-ROMs released in 1995 entitled "Global Data Sets for Land-Atmosphere Models: ISLSCP Initiative 1: 1987–1988." The creation of these CD-ROMs was precipitated by a workshop sponsored by the International Satellite Land Surface Climatology Project (ISLSCP) and held in Columbia, Maryland in June 1992. At this workshop, scientists who model land–atmosphere interactions explained their needs for a comprehensive collection of consistently gridded global data sets of numerous land and atmospheric variables. Further considerations led to the decision to create such a collection starting with the two-year period 1987–1988 and using, as much as possible, algorithms that were already available. Furthermore, all the data sets would be mapped on the same 1° latitude by 1° longitude global grid. In addition to vegetation data, the set of five CD-ROMs contains data on hydrology, soils, snow, ice, oceans, radiation, clouds, and near-surface meteorology. This initial set of CD-ROMs is intended to be followed sometime in 1997 by a new set subtitled "ISLSCP Initiative 2: 1986–1995." The new set will incorporate improvements in the algorithms and provide a finer spatial resolution (0.5° latitude by 0.5° longitude) as well as expanding the temporal coverage to include 10 years of data.

The NDVI vegetation data set provided on the ISLSCP Initiative 1 CD-ROMs is based on the continent-by-continent gridded monthly NDVI values derived from AVHRR data by the Global Inventory Monitoring and Modeling Studies (GIMMS) group at NASA Goddard Space Flight Center. For the ISLSCP CD-ROMs, however, the GIMMS data were adjusted for degradation of the AVHRR instrument over the course of the data record and were regridded onto the 1° latitude by 1° longitude global grid used on the CD-ROMs. The NDVI data in turn were used as a principal input in the calculation of other vegetation-related parameters on the CD-ROMs, such as the "greenness" of the vegetation, a "leaf area index," the land surface albedo (reflectivity), and estimates of basic land-cover or vegetation types.

Review Questions

1. It is mentioned in the text that the Landsat Thematic Mapper, the Landsat Multispectral Scanner, the NOAA AVHRR, and the DMSP SSMI have all been used for vegetation studies, at spatial resolutions of about 30 meters, 80 meters, 1.1 kilometers, and 30 kilometers, respectively. On the basis of these resolutions:
   a. Which one of the four instruments can obtain the greatest detail at a local level?
   b. Which two of the instruments can obtain near-global data coverage with the fewest data points?

2. Why was the AVHRR instrument changed in 1979, narrowing the 0.55–0.90 micrometer channel to 0.58–0.68 micrometers?

3. What does the acronym NDVI stand for?

4. **a.** What advantage does vegetation gain in absorbing red radiation?
   b. What advantage does vegetation gain in reflecting near-infrared radiation?

5. **a.** When vegetation absorbs red radiation, does this process tend to increase or to decrease the value of R measured by the satellite AVHRR instrument? Why?
   b. When vegetation reflects near-infrared radiation, does this process tend to increase or to decrease the value of I measured by the satellite AVHRR instrument?
   c. Does the tendency of vegetation to absorb red radiation and reflect near-infrared radiation lead to high values of I – R or to low values of I – R?

6. Name some instances when the calculated NDVI value would not be a good measure of vegetation coverage.

7. How is the thermal-infrared channel, at 11.5–12.5 micrometers, used in the NDVI algorithm?

8. Mapped monthly NDVI values are generally maximum NDVI values over the course of the month instead of monthly average values. Why?

9. Name three ways in which the ISLSCP Initiative 2 CD-ROM data sets under preparation in 1996 are intended to improve upon the ISLSCP Initiative 1 data sets released in 1995.

Satellite Images of Land Vegetation

Figures 7.6 and 7.7 provide global images of NDVI monthly values for January, April, July, and October 1987, showing the basic seasonal cycle for one year. (New Zealand is a special case in that the New Zealand values in Figures 7.6 and 7.7 are not specific to 1987 but are the values determined from the entire 1982–1990 period, for each of the respective months. The reason for this complication is that New Zealand was not included in the continent-by-continent GIMMS data set used as the base for creating the global images.) Missing data are indicated by the orange coloring at NDVI = 0.0; land locations without missing data have NDVI values of at least 0.05 and no greater than 0.65; oceans, which often have a negative NDVI, are colored white. Deserts display amongst the lowest of the land values and densely vegetated regions amongst the highest.

Missing data are indicated on Figures 7.6 and 7.7 in Antarctica and most of Greenland throughout the year, at the southern tip of South

January 1987

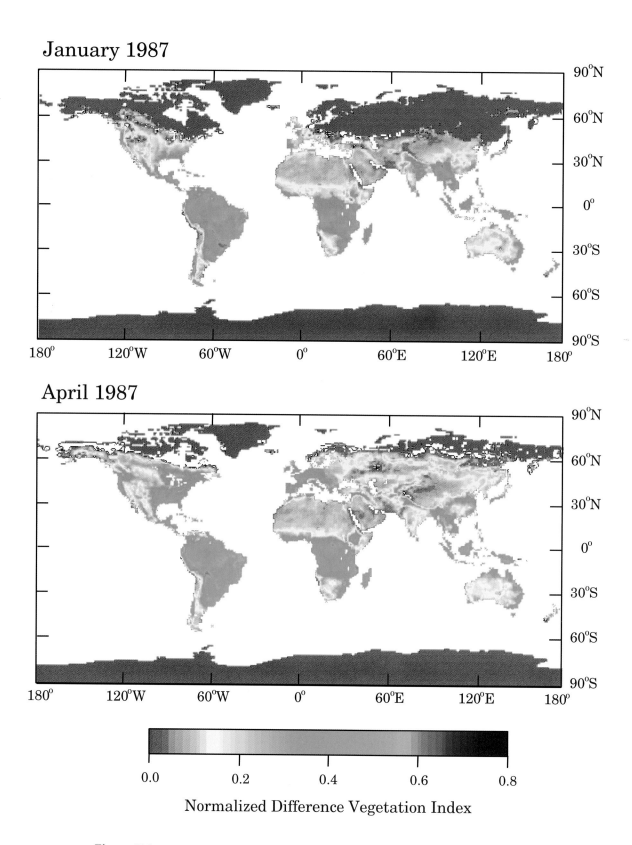

April 1987

0.0   0.2   0.4   0.6   0.8

Normalized Difference Vegetation Index

Figure 7.6.
Global maps of monthly values of the Normalized Difference Vegetation Index
(NDVI) for January and April 1987, calculated from the data of the NOAA 9
AVHRR. The data were obtained from the ISLSCP Initiative 1 CD-ROM.

July 1987

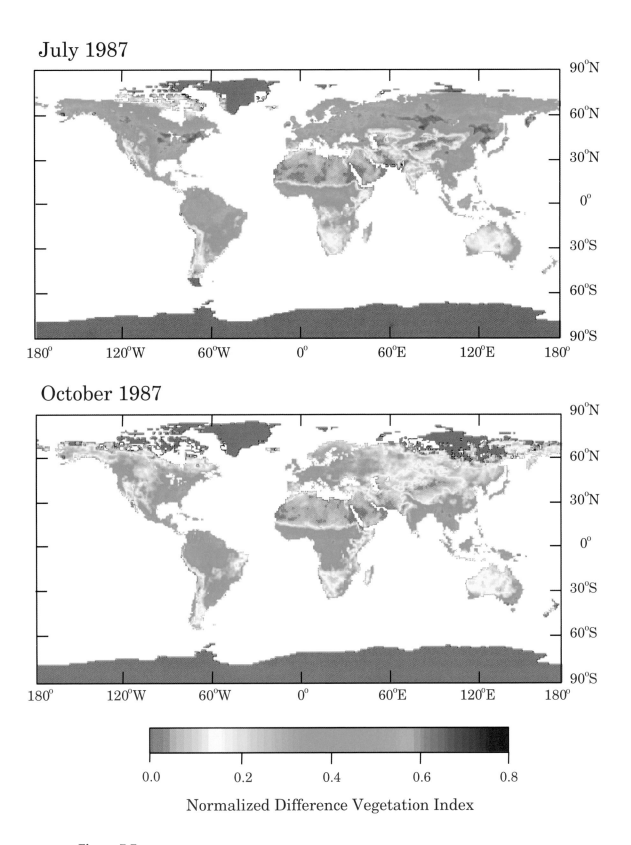

October 1987

Figure 7.7.
Global maps of monthly values of the Normalized Difference Vegetation Index (NDVI) for July and October 1987, calculated from the data of the NOAA 9 AVHRR. The data were obtained from the ISLSCP Initiative 1 CD-ROM.

113

America in July, and over sizable areas of the high latitudes of North America and Eurasia in April, October, and especially January. Two primary reasons account for these missing data: snow cover, making the NDVI inappropriate, and the technique used for screening out cloud-obscured data (described in the previous section), which also eliminates data from areas where the ground surface is particularly cold.

Rather than providing exact quantitative measures of vegetation, the NDVI images are intended to indicate only the general patterns of vegetation distribution and basic seasonal changes in vegetation or photosynthetic activity. For instance, over mid-latitude (30°N–60°N) North America and Eurasia, the values are generally lowest in winter (e.g., January), when much of the vegetation is dormant, and increase markedly in spring and summer, with the blossoming of seasonal vegetation (Figures 7.6 and 7.7). Decreases in NDVI over mid-latitude North America and Eurasia from July to October occur partly from harvesting of crops and partly from crops ripening and leaves being shed.

To illustrate interannual differences, Figures 7.8 and 7.9 provide the 1988 versions of the 1987 images of Figures 7.6 and 7.7. For each month shown, the differences between the two years are quite small compared to the seasonal differences within either year. Still, some interannual differences can be readily identified, such as the lower NDVI values in southern Africa in April 1987 than in April 1988 and the higher NDVI values in France in October 1987 than in October 1988 (Figures 7.6–7.9).

In keeping with the global and hemispheric-scale emphasis of this book, the images presented in this section are AVHRR global maps. However, the interested reader is encouraged to read the 1993 article in the journal *Science* by David Skole and Compton Tucker (listed under "Additional Reading") for a more detailed study that uses 1978 and 1988 Landsat data to map deforestation in the Amazon Basin of Brazil to a much higher spatial resolution than is used in the maps of Figures 7.6–7.9. The study by Skole and Tucker shows the deforested area in Brazil to have risen from 78,000 square kilometers in 1978 to 230,000 square kilometers in 1988 and estimates the much greater spatial extent of the impact of this deforestation on biological diversity. Fortunately, the rate of deforestation in Brazil subsequently decreased, following the 1988 decision by the Brazilian government to rescind laws that had encouraged forest removal.

The vegetation data presented in Figures 7.6–7.9, along with data for each of the other months of 1987 and 1988, are available on volume 1 of the five-volume CD-ROM set "Global Data Sets for Land-Atmosphere Models. ISLSCP Initiative 1: 1987–1988," by B. W. Meeson, F. E. Corprew, J. M. P. McManus, D. M. Myers, J. W. Closs, K.-J. Sun, D. J. Sunday, and P. J. Sellers, NASA Goddard DAAC Science Data Series, 1995. This CD-ROM set is available from the NASA Goddard Distributed Active Archive Center (DAAC), Greenbelt, Maryland and contains data sets for many

January 1988

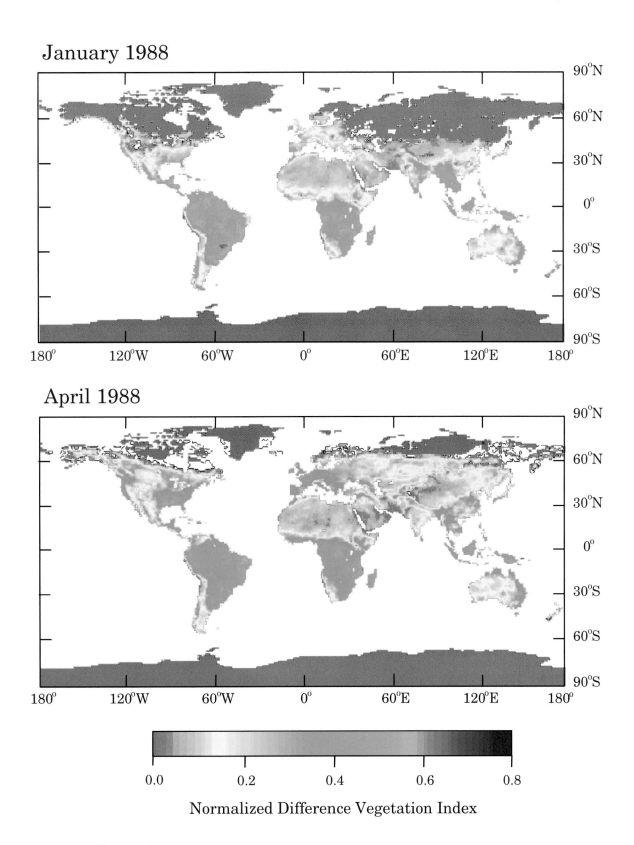

April 1988

Figure 7.8.
Global maps of monthly values of the Normalized Difference Vegetation Index
(NDVI) for January and April 1988, calculated from the data of the NOAA 9
AVHRR. The data were obtained from the ISLSCP Initiative 1 CD-ROM.

115

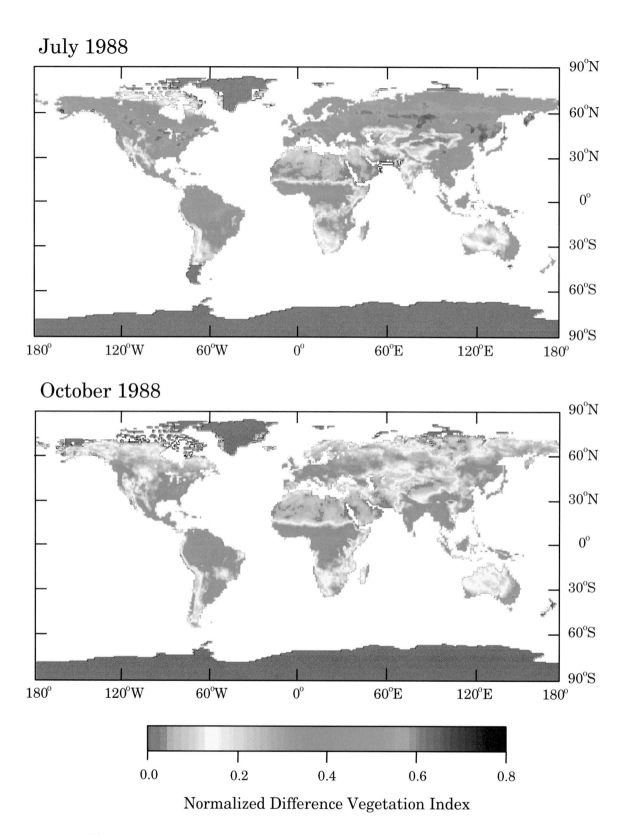

July 1988

90°N
60°N
30°N
0°
30°S
60°S
90°S

180°   120°W   60°W   0°   60°E   120°E   180°

October 1988

90°N
60°N
30°N
0°
30°S
60°S
90°S

180°   120°W   60°W   0°   60°E   120°E   180°

0.0   0.2   0.4   0.6   0.8

Normalized Difference Vegetation Index

Figure 7.9.
Global maps of monthly values of the Normalized Difference Vegetation Index
(NDVI) for July and October 1988, calculated from the data of the NOAA 9
AVHRR. The data were obtained from the ISLSCP Initiative 1 CD-ROM.

116

variables, including NDVI. The NDVI data are described and evaluated in a 1994 article by S. O. Los, C. O. Justice, and C. J. Tucker in the *International Journal of Remote Sensing*, listed under "Additional Reading."

Questions Regarding the Satellite Imagery

1. Using Figure 7.6, in January 1987, which continent has the largest region with NDVI values exceeding 0.4?

2. Of the months January, April, July, and October 1987, which one had:
   a. The highest NDVI values, overall, in North America?
   b. The highest NDVI values, overall, in Eurasia?
   c. The highest NDVI values, overall, in Spain and Portugal?
   d. The highest NDVI values, overall, in India?

3. **a.** From the answers to Questions 2.a and 2.b, overall are the North American and Eurasian NDVI values higher in summer or in winter?
   b. Consider the South American results and summer and winter in the Southern Hemisphere. Overall, were the South American NDVI values in 1987 higher in summer or in winter?
   c. In the Southern Hemisphere portion of Africa, were the 1987 NDVI values greater in the Southern Hemisphere summer or winter?

4. For each of the following months, indicate whether most of the very high NDVI values (greater than 0.5) in Africa are in the Northern Hemisphere or in the Southern Hemisphere:
   a. January 1987.
   b. April 1987.
   c. July 1987.

5. In which direction, northward or southward, did the high NDVI values of tropical Africa tend to move from April to July 1987? This prominent shift in the location of high NDVI values relates to the seasonal movement of the zone of highest tropical precipitation.

6. Relate the changes in the basic patterns of NDVI over Russia in January, April, July, and October 1987 to the seasonal cycle.

7. For each of the following regions, indicate whether the NDVI values, overall, increased or decreased in proceeding from July 1987 to October 1987:
   a. North America.
   b. Brazil.
   c. Western Australia.
   d. Southern Asia.
   e. Central and northern Asia.

8. From the evidence in Figures 7.6 and 7.8, which year, 1987 or 1988, appears to have had the colder January in eastern Europe?

9. Which year, 1987 or 1988, had higher NDVI values in eastern Europe in April?

10. From the evidence in Figures 7.7 and 7.9, do the July and October conditions in the vicinity of the Caspian Sea appear to be more desertlike to the east of the sea or to the west of the sea?

11. Comparing the NDVI maps of Figures 7.6–7.9 with the distribution of deserts and rainforests in Figure 7.1a:
    a. In which seasons are the NDVI values greater in the tropical rainforests of South America and Africa than in the Sahara Desert?
    b. In which seasons are the NDVI values greater in the tropical rainforests of South America and Africa than in the desert of Saudi Arabia?
    c. In which seasons are the NDVI values greater in the tropical rainforests of southeast Asia than in the Gobi Desert of central Asia?

CHAPTER 8

Volcanoes

Introduction

Explosive volcanic eruptions are amongst the most fearsome and spectacular of natural events visible and audible to human observers (e.g., Figure 8.1). Much of a mountain can be blown away within a few hours, and the sound can often be heard across hundreds or even thousands of kilometers. One of the most famous eruptions, that of the Indonesian island of Krakatau (also spelled Krakatoa) on August 27, 1883, obliterated most of the island and was heard approximately 5,000 kilometers (3,100 miles) away, on Rodriguez Island in the Indian Ocean. (Location maps, including the placement of Krakatau midway between Sumatra and Java, are presented in Figure 8.2.)

Volcanoes have a wide variety of impacts, some of which are basically positive from a human perspective and some of which are basically negative. To aid the discussion of a broad selection of these effects, the remainder of this section is divided into subsections on destructive impacts, beneficial impacts, and temperature impacts. Temperature impacts are the most prominent of the broader category of weather impacts, which are generally neither inherently destructive nor inherently beneficial.

Destructive Impacts

In addition to blasting into the atmosphere a substantial volume of material from the erupting or emerging mountain, a volcanic eruption can cause the terrain for kilometers around to be rapidly covered by lava, ash, and pumice, destroying much of the ecosystem over wide expanses and wreaking havoc on human settlements. The blast from Mount Saint Helens in the northwest United States (in the state of Washington) on May

Figure 8.1.
Eruption of Mount Ngauruhoe, New Zealand, January 1974. [Photograph used courtesy of the National Geophysical Data Center, Boulder, Colorado, from the University of Colorado archives.]

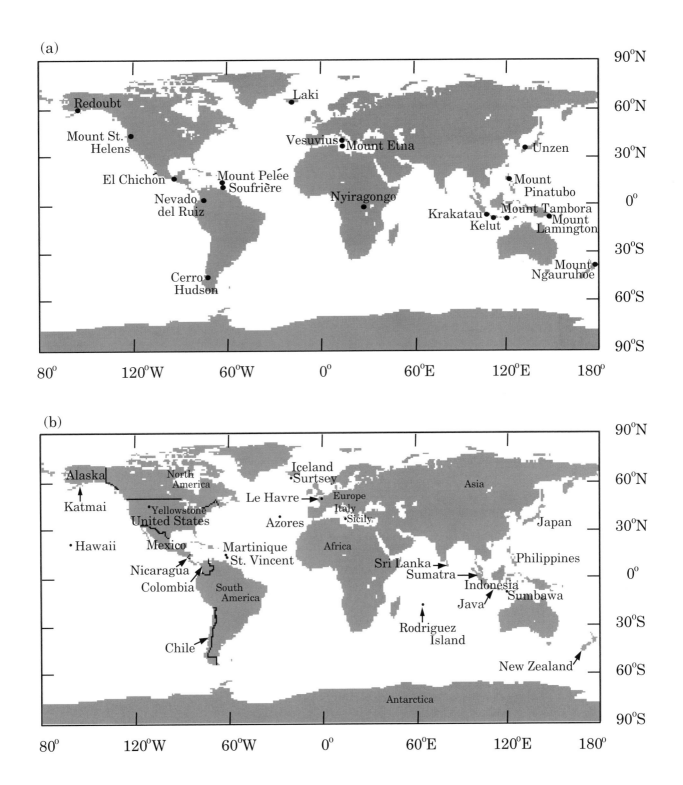

(a)

90°N
60°N
30°N
0°
30°S
60°S
90°S

Redoubt
Mount St. Helens
El Chichón
Nevado del Ruiz
Cerro Hudson
Mount Pelée
Soufrière
Laki
Vesuvius
Mount Etna
Nyiragongo
Unzen
Mount Pinatubo
Krakatau
Kelut
Mount Tambora
Mount Lamington
Mount Ngauruhoe

80°   120°W   60°W   0°   60°E   120°E   180°

(b)

90°N
60°N
30°N
0°
30°S
60°S
90°S

Alaska
Katmai
Hawaii
North America
Yellowstone
United States
Mexico
Nicaragua
Colombia
South America
Chile
Martinique
St. Vincent
Azores
Le Havre
Iceland
Surtsey
Europe
Italy
Sicily
Africa
Sri Lanka
Sumatra
Indonesia
Java
Rodriguez Island
Asia
Japan
Philippines
Sumbawa
New Zealand
Antarctica

80°   120°W   60°W   0°   60°E   120°E   180°

Figure 8.2.
Location maps. (a) Locations of the volcanoes mentioned in the text. (b) Other locations mentioned in the text and questions.

Figure 8.3.
Mount Saint Helens nine years after the May 1980 eruption. A portion of the devastation wrought by the eruption is still clearly visible in the mountain's covering of downed trees. [Photograph from the author.]

18, 1980 flattened millions of trees over nearly 600 square kilometers and destroyed much other plant and animal life as well (Figure 8.3). In one hour, a major pyroclastic flow consisting of pumice, dust, and gas can expand outward 100 kilometers from its source, depositing pumice over thousands of square kilometers to a thickness of several meters. At such rates and thicknesses, most plant and animal life and many human constructs have little chance of survival.

Because of their mobility, animals have a greater chance of survival than plants, and because of forewarnings and evacuations, humans often have a greater chance than many other animal species. Nonetheless, human death tolls from violent eruptions can number in the tens of thousands (e.g., Table 8.1). In fact, such tolls can occur within minutes, as on May 8, 1902 in Saint Pierre on the island of Martinique, where an estimated 30,000 people died in less than three minutes when a cloud of hot dust and gas reached the town following the eruption of Mount Pelée. There were only two known survivors in Saint Pierre, one a jailed convict partially protected in an underground cell. The largest eruption in recorded history, that of Mount Tambora on the Indonesian island of Sumbawa on April 10–11, 1815, killed all but 26 of the 12,000 inhabitants of Sumbawa and, by some estimates, perhaps as many as 90,000 people altogether (Table 8.1).

Massively destructive agents can travel through the water as well as through the atmosphere and on the ground. Specifically, giant waves created by an eruption sometimes rise 40 or more meters above the surface of the ocean and wash away entire communities as they plunge onto shorelines. Such waves, termed *tsunamis* (a term that also includes waves generated by earthquakes), can contribute significantly to the death toll caused by the eruption. The tsunamis fanning out from Krakatau killed approximately 36,000 people on Java and Sumatra (Table 8.1). Forty meters (120 feet) high in the vicinity of Krakatau, the tsunamis lost height but remained recognizable as they spread through much of the world's oceans over the next two days. They were reduced to about 1 meter in height by the time they arrived at Sri Lanka and about 1 centimeter in height when they arrived at Le Havre, France, 32 hours after the explosion.

Other potentially deadly and financially devastating effects of volcanoes include debris avalanches, landslides, mudflows, earthquakes, inundation of population centers with lava flows, and aircraft hazards from volcanic plumes. A mudflow generated by the November 13, 1985 eruption of Nevado del Ruiz in Colombia wiped out the town of Armero and killed more than 22,000 people (Table 8.1). Earthquakes from the 1815 Tambora eruption were felt as far away as 600 kilometers, on Java, and ash fallout from the same eruption extended at least 1300 kilometers from the volcano. Although not a concern at the time of pre-twentieth-century eruptions, volcanic ash from an eruption plume can now also be a major hazard to air travel, as its intake can cause the shutdown of modern jet engines, largely because the engines' operating temperatures exceed the melting point of volcanic materials. Aircraft hazards can extend thousands of kilometers from the volcano vent and could cost billions of dollars and hundreds of lives if a crash were to ensue as a result of a volcanic plume. One noted near-tragedy occurred on December 15, 1989 as all four engines on a Boeing 747-400 aircraft flamed out after the aircraft entered a volcanic ash cloud from Redoubt, an Alaskan volcano 175 kilometers southwest of Anchorage and 285 kilometers from the unfortunate aircraft. The plane descended almost 4 kilometers, from an initial altitude of about 7.6 kilometers, before the crew was able to restart the engines. Although a crash was averted, damage to the aircraft resulted in $80 million in repairs.

In view of the variety of impacts and the distances affected by a major eruption, the danger of damage from volcanoes will probably never be eliminated. Fortunately, many major eruptions are preceded for several

Table 8.1
Death tolls from selected noted eruptions

Volcano	Eruption Date	Estimated Deaths	Primary Causes of Death
Mount Vesuvius, Italy	August 24–25, 79	16,000	ash, pumice, mud, lava
Mount Etna, Sicily	March 25, 1669	20,000	lava
Mount Tambora, Indonesia	April 10–11, 1815	12,000 (direct) 90,000 (total)	dust and ash starvation, after crop and livestock destruction
Krakatau, Indonesia	August 27, 1883	36,000	tsunamis
Soufrière, St. Vincent, British West Indies	May 7, 1902	2,000	cloud of hot gas
Mount Pelée, Martinique, French West Indies	May 8, 1902	30,000	cloud of hot gas and dust
Kelut, Java, Indonesia	May 19–20, 1919	5,000	mudflows
Mount Lamington, Papua New Guinea	January 21, 1952	3,000	cloud of hot gas
Nyiragongo, Zaire	January 10, 1977	70	hot lava
Mount St. Helens, United States	May 18, 1980	60	pyroclastic blast, flying objects, inhalation of hot dust
El Chichón, Mexico	March 28– April 4, 1982	2,000	pyroclastic surges
Nevado del Ruiz, Colombia	November 13, 1985	22,000	mudflows
Mount Pinatubo, Philippines	June 15, 1991	300–900	pumice, ash, gases
Unzen, Japan	May/June, 1991	40	cloud of hot gas

weeks by abundant geologic warnings, such as small earthquakes, gas emissions, and minor eruptions. Death tolls and other damage can be vastly reduced if the warnings are heeded. In the case of the June 15, 1991 eruption of Mount Pinatubo in the Philippines, the death toll of 300–900 people is only a small fraction of what it might have been if 200,000 people had not been evacuated during the days preceding the eruption.

Beneficial Impacts

Although the deaths and significant ecological and property damage caused by eruptions naturally capture considerable attention, volcanic eruptions also have overwhelmingly positive impacts both for the Earth as a whole and for its human inhabitants. The gases released from the Earth's interior through eruptive activity are believed to have formed the Earth's atmosphere billions of years ago and still continue to modify it. Also, the fertility of the soils in the vicinity of a volcano (Figure 8.4) can

Figure 8.4.
Lush agricultural fields on volcanic soils on Faial Island in the Azores.
[Photograph kindly provided courtesy of Jim Garvin.]

be markedly increased by volcanic outpourings, which contain such important nutrients as phosphorus, potassium, calcium, magnesium, and sulfur. At any given time, some of the Earth's most fertile soils are ones that have been overlain by volcanic ash within the previous few months or years or by lava within the previous few decades or centuries.

Other benefits from volcanoes include the following:

(a) New islands can form as underwater eruptions burst through the sea surface. For example, Surtsey, Iceland emerged and grew over the period November 4, 1963 to May 1965, and in the much more distant past, the rest of Iceland, all the Hawaiian Islands, all the Philippine Islands, and hundreds of others emerged through volcanic eruptions.

(b) Hot magma provides energy, which is sometimes captured for human uses. For instance, as of the late twentieth century, 7% of New Zealand's electric power is provided by natural steam of volcanic origin, half of Nicaragua's electric generating capacity is from thermal generating plants, and most of the homes in Iceland benefit from the magmatic heating of underground water.

(c) Some ignimbrite rocks, left by pumice flows, make exceptional building materials, being lightweight, good insulators, strong, and yet soft for easy cutting.

(d) Obsidian, a black, pure glass found in some lavas, has sharp cutting edges and is sometimes stunningly attractive, two properties that have led to its use since prehistoric times in tools and jewelry, respectively. Its versatility as a material for tools is attested to by its current use in modern surgery.

(e) Spectacular sunsets often occur in the months following a major eruption, aided by volcanic aerosols and their reflective properties.

(f) Volcanic regions provide the settings for several scenic national parks, including Katmai in Alaska and Yellowstone in Wyoming, Montana, and Idaho (northwest continental United States).

(g) Finally, horrifying as it may be, benefits have even been derived from the volcanic destruction of whole communities, a prime example being the knowledge obtained as a result of the destruction of the ancient Roman towns of Pompeii and Herculaneum following the eruption of nearby Mount Vesuvius on August 24–25, 79 A.D. Pompeii was left buried under nearly 3 meters of volcanic debris and Herculaneum under 20 meters; 16,000 people died, and both towns were abandoned by those who survived. The burials in the volcanic debris preserved remarkable details about life in the first century, to be uncovered by archaeologists and others in the nineteenth and twentieth centuries.

Temperature Impacts

Another major category of impacts of volcanic eruptions, neither inherently detrimental nor inherently beneficial, is the category consisting of the effects on the Earth's weather. These include local effects caused by changes in the surface albedo that result from coverings of lava, ash, or pumice and precipitation effects that result from the addition of particulate matter to the atmosphere. The weather effects most frequently mentioned, however, are those that the materials injected into the atmosphere have on atmospheric temperatures. These effects can last several years and can extend great distances from the eruption site.

Materials injected only into the lower atmosphere, the troposphere, typically settle to the ground within several weeks, as do the heaviest materials injected into the stratosphere. When large quantities of material reach the stratosphere, however, some of the lighter-weight debris can circle the globe and remain in the upper atmosphere for several years. Once in the stratosphere, the added dust, ash, water vapor (H_2O), sulfur dioxide (SO_2), and other gases, along with some of the particles they are converted into, especially sulfuric acid (H_2SO_4, sulfate) aerosols, have a tendency to absorb solar radiation or reflect it back to space, rather than transmitting it through to lower atmospheric levels. This results in less solar radiation reaching the underlying troposphere, thereby contributing toward cooling the troposphere throughout the latitude zone affected. At the same time, the increased absorption of radiation in the stratosphere contributes toward warming that atmospheric layer. Sulfuric acid in particular is a strong absorber of solar radiation and is often created in large quantities following an eruption, through chemical reactions involving sulfur dioxide (SO_2), water vapor (H_2O), and hydroxyl radicals (OH). Much of the sulfur dioxide from an eruption gets converted into sulfuric acid aerosols within a few months, and these aerosols can remain in the stratosphere for several years, warming the stratosphere and contributing toward cooling the underlying troposphere.

In the case of the August 27, 1883 eruption of Krakatau, the plume of volcanic debris circled the globe (in the upper atmosphere) in two weeks; it arrived over southern Africa on August 28, over the west coast of South America on September 2, and back over Indonesia on September 9, moving in the expected east-to-west direction in accord with the general circulation patterns of the tropical stratosphere. The average near-surface cooling in the Northern Hemisphere as a whole for the second year after the eruption, which is the approximate time frame during which eruption-induced cooling is expected to be greatest, has been estimated to be 0.25°C.

The impact of an eruption on the weather depends not only on the size of the eruption but also on several other variables: its location; the specifics of the emitted materials—particularly the amount of sulfurous gas; and whether the eruptive force is predominantly vertical, helping to send the erupted materials into the stratosphere, or predominantly lateral. For instance, in spite of the violence of the eruption, relatively few of the emissions from Mount Saint Helens's May 1980 eruption reached the stratosphere, largely because the blast was directed laterally. In addition, much of the emitted material was ash, which settled out of the atmosphere fairly quickly, rather than sulfur. Hence the effects on the weather outside the immediate vicinity were small, although the blast flattened millions of trees in the vicinity (e.g., Figure 8.3) and generated an ash cloud significant enough to produce measurable ash accumulations (exceeding 1 centimeter in thickness) in Yakima, Washington, 120 kilometers downwind from the volcano.

In contrast to the case of Mount Saint Helens, the eruptions of El Chichón on March 28, April 3, and April 4, 1982, were nearly vertical and sent considerable sulfur-rich material (with a high potential for conversion into sulfuric acid, which absorbs and scatters solar radiation) into the stratosphere. This eruption consequently had a much greater effect on large-scale weather conditions. The eruption cloud from El Chichón traveled westward around the globe in 3–4 weeks, returning approximately to the longitude of El Chichón on April 25; but it remained confined during this period mostly to the Northern Hemisphere low latitudes. Within a year, the cloud had expanded both northward and southward to cover almost the entire Northern Hemisphere and a large part of the Southern Hemisphere.

At polar latitudes, even a vertically explosive, sulfur-rich eruption can fail to have more than a limited weather effect. This is because the atmospheric circulation at those latitudes often causes the erupted materials to circle the globe only in the polar regions. A low-latitude volcano is more likely to have a weather effect extending over a larger area, both because of the longer distance around the globe at a low versus high latitude and because of the greater amount of mixing between low-latitude and mid-latitude air than between mid-latitude and high-latitude air.

The first known speculation explicitly associating a volcanic eruption with subsequent weather was made by Benjamin Franklin, the multitalented American scientist, statesman, and communicator of the eighteenth century. Soon after the 1783–84 winter, Franklin suggested that a possible cause of the unusually cold winter weather in Europe that year, as well as the persistent fog, might have been the 1783 eruption of the Icelandic volcano Laki. Although Franklin did not develop the thought (and also mentioned the alternative possibility of an extraterrestrial cause), eruptions have since become widely recognized to indeed have such effects. Interestingly, the Laki eruption, which lasted eight months, was largely one of lava flows rather than one producing an explosive eruption cloud, which is more commonly associated with widespread weather effects.

The much larger 1815 eruption of Mount Tambora offered supportive evidence for Franklin's speculations, as the summer of the following year was so cold in Europe and North America that 1816 became known as "the year without a summer." The Tambora eruption is estimated to have injected approximately 180 cubic kilometers of material into the atmosphere (approximately nine times the estimates for Krakatau) and to have produced a plume approximately 43 kilometers high.

More recently, the 1991 eruption of Mount Pinatubo is estimated to have injected approximately 20 million tons of sulfur dioxide into the atmosphere and has been widely touted as the reason that global temperatures were lower in 1992 and 1993 than had been forecast prior to the eruption.

Of course, in no case of suspected eruption-induced cooling is it possible to determine exactly what the temperature would have been if the eruption had not occurred. Hence, estimating the magnitude of any cooling effect remains fundamentally speculative. Nonetheless, estimates can now be made with considerable scientific backing, through the use of general circulation models (GCMs) of the atmosphere. Specifically, the cooling can be estimated by generating two simulations using the same GCM, numerically inserting the volcanic debris in one of the simulations and leaving it out in the other, then subtracting the resulting temperatures. Still, some uncertainty will always remain, as the models can never simulate the atmosphere precisely.

Another suspected effect, this time very specific, of a major eruption is the particularly large Antarctic ozone hole in 1993, which might have been related to the 1991 Mount Pinatubo eruption. Volcanic aerosols in the upper atmosphere can trigger chemical reactions that contribute to the destruction of the ozone layer, and this process might well have been a factor in the 1993 case. As with the cooling issue, however, no one can know with certainty exactly how big or deep the 1993 ozone hole would have been if the eruption had not taken place, although numerical simulations can provide sophisticated estimates.

Review Questions

1. How far away can a major eruption sometimes be heard?

2. A volcanic eruption can deposit pumice over an area of thousands of square kilometers. Of the following choices, which would be most reasonable for the average thickness of such a deposit in the case of a major twentieth-century pyroclastic flow: several millimeters, several centimeters, several meters, or several kilometers?

3. What is a tsunami?

4. How can a volcanic eruption be a hazard to jet aircraft even days after the eruption has ended?

5. How important are volcanoes to the Earth's atmosphere?

6. What are some of the nutrients contained in volcanic outflows?

7. Name three benefits from volcanoes in addition to the increased soil fertility resulting from volcanic outpourings.

8. a. How soon do volcanic materials injected into the troposphere (the lower atmosphere) tend to settle to the ground?
   b. How long can sizable amounts of debris resulting from an eruption remain in the stratosphere?

9. a. Is a major volcanic eruption with significant inputs to the upper atmosphere expected to cause warming or to cause cooling of the troposphere? Why?
   b. Is a major volcanic eruption with significant inputs to the upper atmosphere expected to cause warming or to cause cooling of the stratosphere? Why?

10. a. In what time frame after a major sulfur-producing eruption is the widescale near-surface cooling effect expected to be the greatest?
    b. Why is this cooling effect not greatest within the first few days after the eruption?

11. Some eruptions have much greater effects on the weather of the subsequent few years than others. For each of the following pairs, indicate which condition is more likely to lead to greater widespread weather effects:
    a. A vertically explosive eruption or a horizontally explosive eruption.
    b. A sulfur-rich eruption or a sulfur-poor eruption.
    c. A high-latitude eruption or a low-latitude eruption.

12. Why did the 1982 eruption of El Chichón have a much greater effect on large-scale weather than the 1980 eruption of Mount Saint Helens?

13. Who was the eighteenth-century American to make the first known speculation explicitly associating a volcanic eruption with the weather?

14. The year 1816 has been labeled "the year without a summer" because of the low summertime temperatures in Europe and North America. How has Mount Tambora, in Indonesia, been connected to this event?

15. Some scientists had predicted higher temperatures in 1992 and 1993 than actually occurred. What 1991 volcanic eruption is often mentioned as the probable cause of the cooler-than-expected conditions in 1992 and 1993?

16. How can general circulation models of the atmosphere be used to estimate the amount of tropospheric cooling resulting from a volcanic eruption?

Satellite Detection of Volcanic Emissions

Many satellite instruments have provided information about volcanoes. The Landsat Multispectral Scanner (MSS) and Thematic Mapper (TM) have provided images of such activity as volcanic venting, volcanic ash falls, and volcanic lava flows at spatial resolutions of 80 meters and 30 meters for the MSS and TM, respectively. Advanced Very High Resolution Radiometers (AVHRRs) have been used to monitor the dispersal of volcanic plumes and the distribution of lava flows at the coarser resolution of about 1.1 kilometers. Satellite Synthetic Aperture Radars (SARs) have been used to map the distribution of cinder cones, fault zones, and lava flows, and new radar interferometry techniques are being used to monitor the subtle signs of deformation associated with impending eruptions. The Microwave Limb Sounder (MLS) on NASA's Upper Atmosphere Research Satellite (UARS) and the Total Ozone Mapping Spectrometer (TOMS) on NASA's Nimbus 7 and the Russian Meteor 3 spacecraft have been used to measure sulfur dioxide emissions and the spread of sulfur dioxide away from volcanoes and sometimes around the globe. The Stratospheric Aerosol and Gas Experiment (SAGE) on NASA's Earth Radiation Budget Satellite (ERBS) has been used to measure the mass of the sulfuric acid aerosols generated by volcanoes.

In the next section, sample images are presented from the TOMS instrument, which has been the most widely used satellite instrument for examining the atmospheric effects of volcanic eruptions. In particular, the images are from the Nimbus 7 TOMS, which was launched in October 1978 and collected data until May 9, 1993. Global data were collected daily, at a spatial resolution of about 50 kilometers.

The TOMS instrument measures solar ultraviolet radiation that has entered the Earth's atmosphere and been reflected back toward space. It was developed to determine ozone amounts, as discussed in Chapter 3. After the April 1982 eruption of El Chichón, however, the calculated TOMS ozone values were found to be abnormally high over Mexico, and it was soon determined that the reason for the erroneous ozone results was that reduced ultraviolet amounts were being interpreted as due to ozone when in reality they were due to sulfur dioxide from the El Chichón eruption. Once this was recognized, the TOMS ozone algorithm was adjusted to correct for the presence of sulfur dioxide, and calculations to obtain sulfur dioxide were added. Similarly to the ozone calculations, the sulfur dioxide calculations are based on the amounts of reflected radiation received by the satellite instrument at various ultraviolet wavelengths and the known properties of sulfur dioxide regarding absorption at those wavelengths. In the case of sulfur dioxide, the calculations are based on data from four channels on the TOMS, those at 0.3125, 0.3175, 0.3312, and 0.3398 micrometers. For the first two of these wavelengths, sulfur dioxide is a much stronger absorber of radiation than ozone is, whereas for the last two, ozone is the stronger absorber. Hence, high levels of atmospheric sulfur dioxide result in a relatively greater depression of the radiation values at wavelengths of 0.3125 micrometers and 0.3175 micrometers and a relatively lesser depression of the radiation values at wavelengths of 0.3312 micrometers and 0.3398 micrometers than do high levels of atmospheric ozone, helping researchers to distinguish the signals from the two gases.

The accuracy of the TOMS sulfur dioxide results varies depending upon the circumstances. Some of the complicating factors lessening the accuracies are volcanic ash, aerosols, low Sun angle, heavy cloud cover, significant dispersal of the plume before the measurements are taken, and conversion of the sulfur dioxide to sulfuric acid (which is not detected by the TOMS in the way sulfur dioxide is) before the measurements are taken. Overall, the sulfur dioxide measurements have been estimated to be accurate to within about $\pm 30\%$, although for the best conditions for the TOMS, that is, low-latitude volcanoes and little or no ash content, the accuracy has been estimated to be improved to ± 5–10%.

The TOMS data have been used not just to map and track volcanic plumes but also to calculate total sulfur dioxide emissions. The calculation of total SO_2 emissions is done by integrating the atmospheric column amounts (in units indicating the thickness of the SO_2 gas) over the area affected, obtaining the volume of sulfur dioxide, and then multiplying the result by the density of sulfur dioxide to obtain the mass. For the two largest eruptions that occurred during the operation of the Nimbus 7 TOMS, estimates from the TOMS data indicate that El Chichón sent approximately 7 million tons of sulfur dioxide into the atmosphere and Mount Pinatubo sent approximately 20 million tons of sulfur dioxide into the atmosphere.

The fact that volcanic plumes are identified through TOMS data by virtue of their sulfur dioxide content should not be misinterpreted as meaning that sulfur dioxide is the primary emitted gas. Water vapor and carbon dioxide are generally emitted from a volcano in much greater quantities than sulfur dioxide is. Water vapor and carbon dioxide, however, are both sufficiently common in the atmosphere that tracking a volcanic cloud using either of those two emissions would be difficult. In contrast, the sulfur dioxide concentration in a volcanic plume is often 10 to 100 times greater than in the background atmosphere and is comparatively easy to detect and measure.

Review Questions

1. In addition to the Landsat Multispectral Scanner and Thematic Mapper, the text mentions the following five satellite instruments as having been used to obtain information about volcanoes: the Advanced Very High Resolution Radiometer (AVHRR), the Synthetic Aperture Radar (SAR), the Microwave Limb Sounder (MLS), the Total Ozone Mapping Spectrometer (TOMS), and the Stratospheric Aerosol and Gas Experiment (SAGE). Of these five:
   a. Which two were mentioned as having been used to examine surface features created by volcanoes?
   b. Which four were mentioned as having been used to examine atmospheric phenomena created by volcanoes?

2. Which satellite instrument has been the one most widely used for examining the atmospheric effects of volcanic eruptions?

3. After the El Chichón eruption, the TOMS ozone algorithm obtained some erroneously high ozone values in the vicinity of the eruption plume. What good came out of these errors?

4. a. What region of the electromagnetic spectrum is used by the TOMS instrument to determine sulfur dioxide amounts in the atmosphere?
   b. Why is this region useful for determining sulfur dioxide?

5. Name three complicating factors that, if existent for the particular eruption and measuring conditions, would worsen the accuracy of the TOMS sulfur dioxide results.

6. Once the TOMS data have been used to calculate the thickness of sulfur dioxide gas throughout the spatial expanse of a volcanic plume, how can those results be used to determine the total mass of sulfur dioxide emitted by the volcano (and still within the same plume)?

7. The largest eruption during the operation of the Nimbus 7 TOMS (October 1978 to May 1993) has been calculated to have injected approximately 20 million tons of sulfur dioxide into the atmosphere. Which eruption was this?

8. Volcanoes generally emit much more water vapor and carbon dioxide than sulfur dioxide. Why, then, is a volcanic plume not monitored by tracking its water vapor and carbon dioxide emissions?

Satellite Images of Volcanic Emissions

As mentioned in the previous section, volcanoes have been imaged using data from AVHRR, Landsat, and SAR, all of which have provided important information about volcanoes and their impacts. In this section, volcanic emissions to the atmosphere, rather than the volcanoes themselves, are used for the illustrative examples. Specifically, examples are provided of the use of the Nimbus 7 TOMS to track the volcanic emissions from two eruptions: the June 1991 eruption of Mount Pinatubo mentioned in the previous two sections and the August 1991 eruption of Cerro Hudson in southern Chile. In both cases, the data presented are sulfur dioxide measurements in units of milli atmospheres centimeter, which give the thickness of the layer of pure sulfur dioxide that would be created if all the atmospheric sulfur dioxide were to be isolated from the rest of the atmosphere and brought down to sea level. One milli atmosphere centimeter refers to a thickness of one-thousandth of a centimeter. Although these units are numerically identical to the Dobson units used for plotting ozone data in Chapter 3, scientists studying volcanic emissions generally use the term *milli atmospheres centimeter* (abbreviated milli atm cm) rather than the term *Dobson units*, which is used by scientists studying ozone in honor of the early work by G. M. B. Dobson regarding the stratospheric ozone layer. Atmospheric sulfur dioxide amounts are usually even less than atmospheric ozone amounts, being predominantly less than 15 milli atmospheres centimeter (i.e., the sulfur dioxide thickness through the entire depth of the atmosphere is generally less than 0.015 centimeter). After an eruption, local sulfur dioxide values often increase to 100 milli atmospheres centimeter or more, and after the Mount Pinatubo eruption, some local values temporarily exceeded 600 milli atmospheres centimeter.

Figures 8.5 and 8.6 show the progression of the sulfur dioxide cloud from Mount Pinatubo every three days from June 16 through June 25, 1991, following the cataclysmic eruption on June 15 (which had been preceded by smaller eruptions and considerable volcanic activity over the course of several days). From June 16 to June 19, the cloud moved southwestward from its origin at 15°09′N, 120°19′E, after which, over the next six days, it continued westward, although no longer with a southward component as well (Figures 8.5 and 8.6). Over the full 10-day period, the sulfur dioxide cloud quite noticeably dispersed, spreading out to cover a much larger area, with markedly reduced sulfur dioxide concentrations. Subsequent to the time frame of Figures 8.5 and 8.6, the cloud continued around the globe, returning to the longitude of Mount Pinatubo on about July 7, 22 days after the major eruption. Although peak sulfur dioxide values remained above 200 milli atmospheres centimeter over a considerable area on June 19, reductions in the peak values were rapid over the next

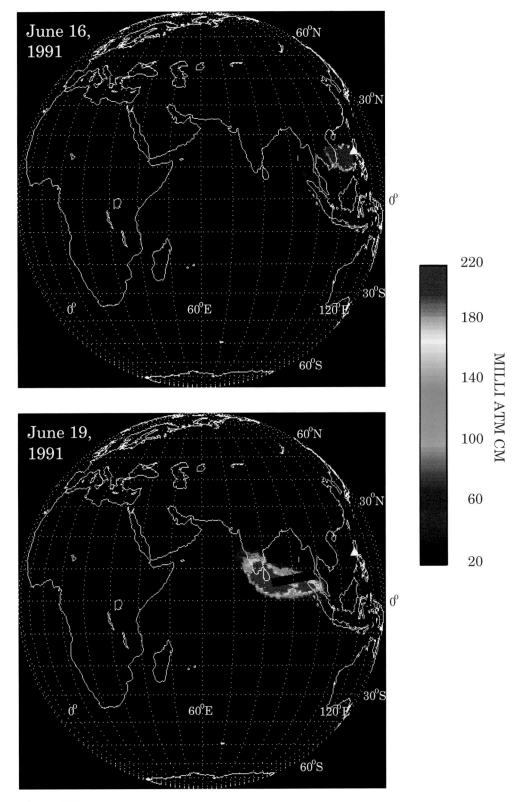

Figure 8.5.
Atmospheric sulfur dioxide amounts on June 16 and June 19, 1991, following the June 15 eruption of Mount Pinatubo, calculated from the data of the Nimbus 7 TOMS. The images were obtained in electronic form from the TOMS SO$_2$ Group at NASA Goddard Space Flight Center, with labels added later. [The prominent rectangular black strip across a portion of the volcanic cloud on the June 19 image is a band of missing data.]

135

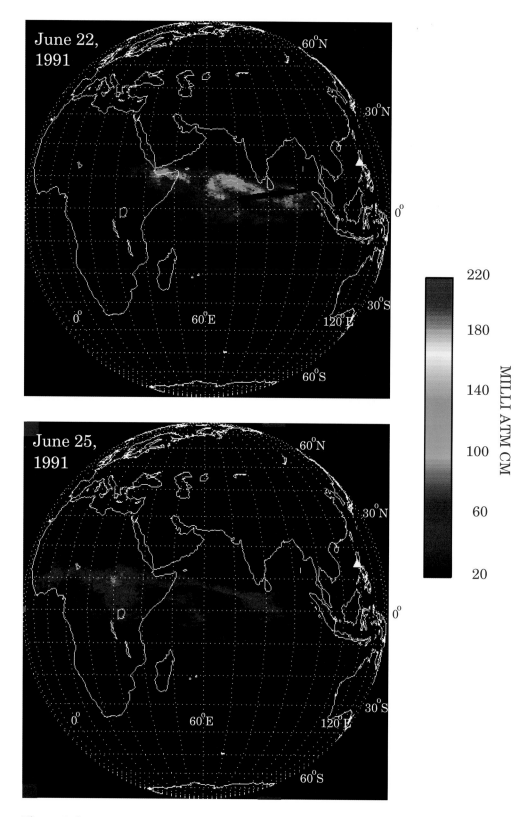

Figure 8.6.
Atmospheric sulfur dioxide amounts on June 22 and June 25, 1991, following the June 15 eruption of Mount Pinatubo, calculated from the data of the Nimbus 7 TOMS. The images were obtained in electronic form from the TOMS SO$_2$ Group at NASA Goddard Space Flight Center, with labels added later. [The prominent rectangular black strip across a portion of the volcanic cloud on the June 22 image is a band of missing data.]

several days (Figures 8.5 and 8.6). These reductions came about partly from dispersal of the cloud but also partly from conversion of sulfur dioxide to sulfuric acid, which is not measured by the TOMS. The sulfuric acid amounts are believed to have spread over large expanses of both hemispheres and to have had a prominent cooling effect during the subsequent two years.

In contrast to Mount Pinatubo, which is located in the low latitudes of the Northern Hemisphere, Cerro Hudson is in the mid-latitudes of the Southern Hemisphere, in the southern Andes mountains of Chile at 45°55'S and 73°0'W (Figure 8.2a). The 1991 Cerro Hudson eruptions began with several days of minor activity, followed by large ash and sulfur dioxide emissions from August 12 through August 15, highlighted by major clouds emitted on August 13 and especially August 15. Figures 8.7 and 8.8 present the daily progression of the Cerro Hudson emission cloud (combining the August 12–15 emissions) from August 15 through August 22. The Cerro Hudson output of sulfur dioxide was roughly twice the sulfur dioxide output of the 1980 eruption of Mount Saint Helens, although only about one tenth of the output of Mount Pinatubo. Because of the sulfur dioxide contrast between Cerro Hudson and Mount Pinatubo, if the same scale were used for Cerro Hudson in Figures 8.7 and 8.8 as for Mount Pinatubo in Figures 8.5 and 8.6, the Cerro Hudson cloud would not show up well. Consequently, the scale used in Figures 8.7 and 8.8 has been changed for the Cerro Hudson images, with the range of values reduced in order to highlight the Cerro Hudson emissions.

One of the consequences of the difference in locations of the two volcanoes is a difference in the direction of movement of the emission clouds. The Mount Pinatubo cloud traveled predominantly westward, while spreading also to the north and south, whereas the Cerro Hudson cloud moved predominantly eastward (Figures 8.7 and 8.8), as would most volcanic clouds at roughly the same altitudes in the mid-latitudes of either hemisphere. In both cases the cloud movements were responding to the basic circulation patterns of the atmosphere. As is clear from Figures 8.7 and 8.8, the Cerro Hudson sulfur dioxide cloud elongated considerably as it evolved, forming a lengthy sinuous feature by August 20. In contrast to the Mount Pinatubo cloud, it did not also widen significantly to the north and south (Figures 8.7 and 8.8 versus Figures 8.5 and 8.6). The evolution of the Cerro Hudson cloud, over the period of Figures 8.7 and 8.8 plus the succeeding week and a half, has provided information that has added substantially to the knowledge of stratospheric wind patterns over the latitude range of 45°S–70°S, at least for the August-September 1991 period. The length of time the Cerro Hudson emissions took to reach the south pole and the limited amounts that eventually arrived there provided confirmation of the basic isolation of the polar stratospheric air during the Southern Hemisphere late-winter period (August-September). This iso-

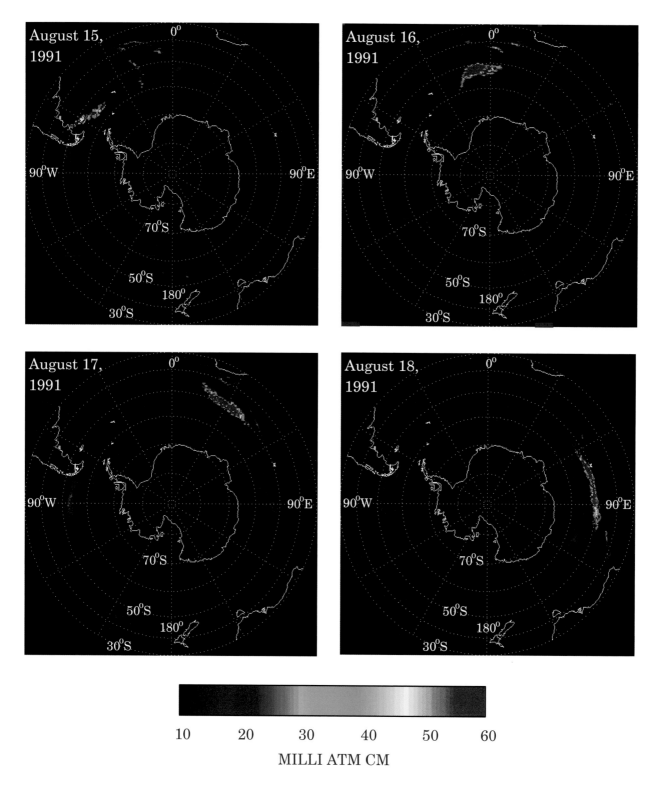

Figure 8.7.
Atmospheric sulfur dioxide amounts on August 15, 16, 17, and 18, 1991, following the August 12–15 eruption of Cerro Hudson, calculated from the data of the Nimbus 7 TOMS. The images were obtained in electronic form from the TOMS SO$_2$ Group at NASA Goddard Space Flight Center, with labels added later.

138

several days (Figures 8.5 and 8.6). These reductions came about partly from dispersal of the cloud but also partly from conversion of sulfur dioxide to sulfuric acid, which is not measured by the TOMS. The sulfuric acid amounts are believed to have spread over large expanses of both hemispheres and to have had a prominent cooling effect during the subsequent two years.

In contrast to Mount Pinatubo, which is located in the low latitudes of the Northern Hemisphere, Cerro Hudson is in the mid-latitudes of the Southern Hemisphere, in the southern Andes mountains of Chile at 45°55'S and 73°0'W (Figure 8.2a). The 1991 Cerro Hudson eruptions began with several days of minor activity, followed by large ash and sulfur dioxide emissions from August 12 through August 15, highlighted by major clouds emitted on August 13 and especially August 15. Figures 8.7 and 8.8 present the daily progression of the Cerro Hudson emission cloud (combining the August 12–15 emissions) from August 15 through August 22. The Cerro Hudson output of sulfur dioxide was roughly twice the sulfur dioxide output of the 1980 eruption of Mount Saint Helens, although only about one tenth of the output of Mount Pinatubo. Because of the sulfur dioxide contrast between Cerro Hudson and Mount Pinatubo, if the same scale were used for Cerro Hudson in Figures 8.7 and 8.8 as for Mount Pinatubo in Figures 8.5 and 8.6, the Cerro Hudson cloud would not show up well. Consequently, the scale used in Figures 8.7 and 8.8 has been changed for the Cerro Hudson images, with the range of values reduced in order to highlight the Cerro Hudson emissions.

One of the consequences of the difference in locations of the two volcanoes is a difference in the direction of movement of the emission clouds. The Mount Pinatubo cloud traveled predominantly westward, while spreading also to the north and south, whereas the Cerro Hudson cloud moved predominantly eastward (Figures 8.7 and 8.8), as would most volcanic clouds at roughly the same altitudes in the mid-latitudes of either hemisphere. In both cases the cloud movements were responding to the basic circulation patterns of the atmosphere. As is clear from Figures 8.7 and 8.8, the Cerro Hudson sulfur dioxide cloud elongated considerably as it evolved, forming a lengthy sinuous feature by August 20. In contrast to the Mount Pinatubo cloud, it did not also widen significantly to the north and south (Figures 8.7 and 8.8 versus Figures 8.5 and 8.6). The evolution of the Cerro Hudson cloud, over the period of Figures 8.7 and 8.8 plus the succeeding week and a half, has provided information that has added substantially to the knowledge of stratospheric wind patterns over the latitude range of 45°S–70°S, at least for the August-September 1991 period. The length of time the Cerro Hudson emissions took to reach the south pole and the limited amounts that eventually arrived there provided confirmation of the basic isolation of the polar stratospheric air during the Southern Hemisphere late-winter period (August-September). This iso-

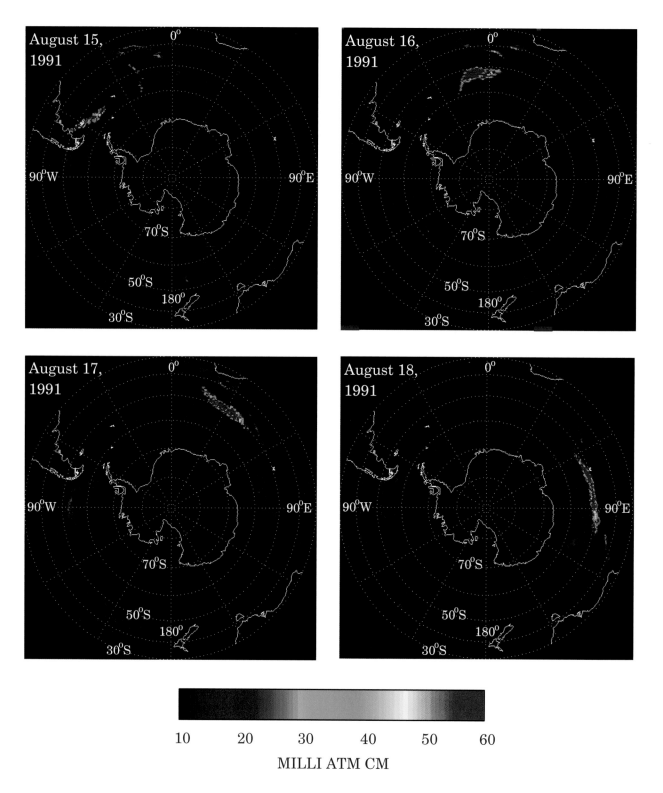

Figure 8.7.
Atmospheric sulfur dioxide amounts on August 15, 16, 17, and 18, 1991, following the August 12–15 eruption of Cerro Hudson, calculated from the data of the Nimbus 7 TOMS. The images were obtained in electronic form from the TOMS SO$_2$ Group at NASA Goddard Space Flight Center, with labels added later.

138

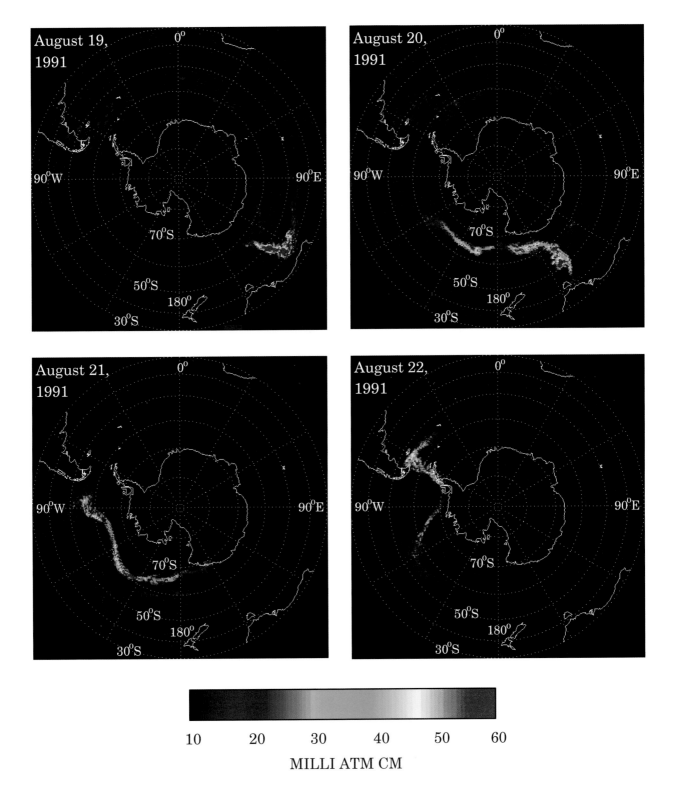

Figure 8.8.
Atmospheric sulfur dioxide amounts on August 19, 20, 21, and 22, 1991, following the August 12–15 eruption of Cerro Hudson, calculated from the data of the Nimbus 7 TOMS. The images were obtained in electronic form from the TOMS SO_2 Group at NASA Goddard Space Flight Center, with labels added later.

139

lation, which continues into early spring, factors into the formation and deepening of the Antarctic ozone hole each September-October (Chapter 3), as it hinders ozone from lower latitudes from reaching the polar region. Another interesting aspect of the Cerro Hudson cloud is how coherent it remained as it circled the globe, retaining peak sulfur dioxide values exceeding 55 milli atmospheres centimeter throughout its initial circuit around the globe (Figures 8.7 and 8.8). Calculations by the TOMS SO_2 Group indicate that the total sulfur dioxide content of the cloud was 1.5 million tons on August 16 and still 1 million tons on August 21.

The TOMS images in Figures 8.5–8.8 were all obtained from the world wide web at the NASA Goddard Space Flight Center TOMS Volcanic SO_2 home page (http://skye.gsfc.nasa.gov/).

Questions Regarding the Satellite Imagery

1. If, after a major volcanic eruption, all of the sulfur dioxide in a vertical column of the atmosphere going straight through the volcanic plume were combined together and brought down to sea level, would the thickness of the sulfur dioxide be closer to the thickness of a paper clip, the height of a newborn infant, the height of a National Basketball Association (NBA) basketball player, the height of a two-story house, or the height of a mountain?

2. About how many weeks did it take debris from Mount Pinatubo to circle the globe (in the upper atmosphere) and return to the longitude of Mount Pinatubo?

3. Which of the images of Figures 8.5 and 8.6 show a portion of the volcanic cloud from Mount Pinatubo having moved into the Southern Hemisphere?

4. For each of the four images in Figures 8.5 and 8.6, indicate which of the following latitude bands contained the majority of the sulfur dioxide cloud: 0°N–10°N, 0°S–10°S, 10°N–20°N, 10°S–20°S, 20°N–30°N, 20°S–30°S.

5. For each of the following dates, what are the approximate highest values of sulfur dioxide amounts indicated on the images of Figures 8.5 and 8.6?
   a. June 16, 1991.
   b. June 19, 1991.
   c. June 22, 1991.
   d. June 25, 1991.

6. Using Figures 8.5 and 8.6, did the leading portion of the volcanic cloud from Mount Pinatubo move westward at a rate that was most comparable to the rate of a human walking, the rate of a human running, the rate of a human driving a car, or the rate of an air-

plane? Assume that the surface for walking, running, or driving is a level land surface and the conditions are comfortable.

7. If the color scale used on Figures 8.7 and 8.8 for the Cerro Hudson eruption were also used for the Mount Pinatubo eruption (Figures 8.5 and 8.6), would there be any red values on the June 25, 1991 image, 10 days after the Mount Pinatubo eruption?

8. For each of the following dates, what are the approximate highest values of sulfur dioxide amounts indicated on the images of Figures 8.7 and 8.8: August 15, August 16, August 17, August 18, August 19, August 20, August 21, and August 22, 1991?

9. **a.** On which of the eight images of Figures 8.7 and 8.8 was the Cerro Hudson sulfur dioxide cloud centered farthest north?
   b. On that image, on approximately what latitude was the cloud centered?

10. From the images of Figures 8.7 and 8.8, did the Cerro Hudson volcanic cloud appear to have encountered more winds from the north and the south over the period August 15–18, 1991 or over the period August 19–22, 1991?

11. **a.** Using Figures 8.7 and 8.8, approximately how many weeks did it take the August 1991 Cerro Hudson volcanic cloud to "circle the globe" (i.e., to complete a circuit through all longitudes and return to its initial longitude)?
    b. Is this longer or shorter than the time it took the Mount Pinatubo cloud to circle the globe?
    c. Which cloud, from the Mount Pinatubo or the Cerro Hudson eruption, had to travel a longer distance to circle the globe?

CHAPTER 9

Conclusions:
Strengths and Limitations
of Satellite Data

The Scope and Utility of Satellite Earth Observations

Over the past three decades, satellite instruments pointed toward the Earth have transmitted a vast storehouse of information about our planet and its surrounding atmosphere. This book provides samples of the resulting satellite images, highlighting six key variables in the Earth's climate system for which satellite data have revealed exciting new global or near-global information—atmospheric ozone, sea ice, continental snow cover, sea surface temperature, land vegetation, and volcanic emissions. Some of the other climate variables now being examined with satellite data are solar radiation reaching the Earth's outer atmosphere, longwave radiation emitted by the Earth, atmospheric temperature, land surface temperature, wind speed, aerosols, water vapor, precipitation, clouds, lightning, several chemical constituents in the atmosphere (in addition to ozone), ocean and land topography, ocean circulation, ocean biological productivity, evaporation from the ocean, the ice sheets of Greenland and Antarctica, and mountain glaciers and ice caps around the world. For all of these variables, satellite instruments are obtaining a record with spatial coverage and temporal detail that would have been inconceivable before the satellite era.

The satellite data yield both immediate practical benefits and longer-term conceptual benefits in the improved understanding of the Earth–atmosphere system. Among the immediate benefits is the tremendous aid they provide to weather forecasters through their depictions of the large-scale weather picture. Weather forecasts are not always accurate, as storm

143

systems can take unexpected twists and turns; but by and large the weather forecasts for the next day or two provide useful indications of the upcoming near-term weather conditions. Furthermore, major storm systems such as hurricanes are no longer likely to arrive along continental coasts without having been spotted earlier and tracked on satellite images (e.g., Figure 1.3). The forewarnings can save hundreds of lives in cases when hurricanes eventually make direct hits on coastal communities.

On a longer time frame, satellite data allow widespread recognition of changes to the landscape that previously could go unnoticed by all except local inhabitants and a few outsiders. A case in point is the shrinking of the Aral Sea in western Asia through much of the twentieth century, for the first half of which no satellite data were available and very few people were aware of the Aral's contraction. The shrinkage resulted largely from the artificial diversion of waters from rivers running into the sea, the purpose being to use the water for irrigating agricultural fields. Although for many years the irrigation was successful and substantially increased yields of cotton and rice, this initial success could not continue, as the fertilizers contaminated the soil and the retreating sea left behind a desert of salt and sand and a host of resulting problems. By monitoring the Earth system, satellites increase the chances that critical changes in the landscape, such as shrinking seas and widespread deforestation (discussed in Chapter 7), will be identified early on and that corrective action will be taken in time to reverse the problems and lessen the damage.

Other environmental issues for which satellite data are providing information include such relatively short-term phenomena as individual forest fires, oil well fires, major oil spills, and the initial dispersal of contaminants from nuclear accidents, as well as the long-term phenomena of the Antarctic ozone hole (discussed in Chapter 3) and the possibility of climate warming or climate cooling. The depiction on satellite imagery of the full extent of an oil spill or a forest fire can help in the efforts to contain the damage, clean up the oil, or stop the fire.

On a more conceptual level, the satellite data sets of the Earth-atmosphere system are helping scientists, philosophers, environmentalists, and others to formulate and quantify emerging, scientifically based wholistic views of the Earth. Most importantly, the satellite data allow observation and monitoring of all regions of the Earth's surface and atmosphere, so that theory and explanations no longer have a database restricted to areas and times where humans have physically been and made the observations or left instruments to record the measurements. For instance, the El Niño (Chapter 6) was recognized as an important phenomenon along the coasts of Peru and Ecuador back in the nineteenth century, but no database existed at that time with which to connect the local occurrences to much larger-scale phenomena. Now, satellite data show the clear connection of occurrences along those coasts to oceanic conditions throughout the equatorial Pacific, enough so in fact that the satellite data can contribute to predictions of the Peru and Ecuador occurrences. Furthermore, by now the El Niño is being further connected conceptually to many phenomena well beyond the equatorial Pacific, including large-scale atmospheric pressure patterns, Indian monsoons, Australian droughts, Californian rainfalls, Atlantic hurricanes, and sea ice in both polar regions. Many of the connections were made without the use of satellite data, but the satellite data are helping in quantifying, testing, and expanding them. Similarly, as the satellite data sets lengthen and the analysis of them deepens, additional examples of the interconnectedness of different locations and different phenomena frequently become apparent. It remains, however, up to people (generally scientists) to determine which apparent connections are significant and which are merely coincidental or sporadic.

The interested reader can find more information on each of these topics and variables, plus on the use of satellites to examine them, in the *Atlas of Satellite Observations Related to Global Change* listed under "Additional Reading."

Review Questions

1. Excluding the six variables highlighted in Chapters 3–8, list six additional variables being examined with satellite data.

2. How do satellites help in the forecasting of the possibility that a hurricane will strike a continental coast?

3. Explain how humans contributed to the shrinking of the Aral Sea during the twentieth century.

4. Name an environmental advantage of having satellite data reveal such phenomena as shrinking seas and expanding deforestation.

5. How can the depiction on satellite imagery of an oil spill help in the clean-up efforts?

6. How are satellite data sets helping in the formulation and quantification of wholistic views of the Earth?

Limitations

Despite the broad range and tremendous value of satellite data, viewing satellite-derived data sets as perfect depicters of the global environment and changes within the environment would be a major mistake. This book has identified several of the many limitations of the data and some of the cautions regarding their use. To reiterate a few of the key examples:

(a) First and foremost, raw satellite data are rarely of the variable that scientists seek to determine. Satellite instruments do not collect samples of vegetation, sea ice, or snow cover. They do not collect air samples to measure ozone or sulfur dioxide amounts, or sea water samples to measure sea surface temperatures. Instead, they simply receive and transmit radiation, although often from very carefully selected regions of the electromagnetic spectrum. The radiation amounts are then analyzed for the information they are thought to contain about the various Earth-system parameters of interest, including vegetation, sea ice, snow cover, atmospheric ozone and sulfur dioxide, and sea surface temperature.

Generally, an algorithm (or set of equations) is used to convert from the radiation values to the desired geophysical parameters. Because humans create the algorithms and interpret the data products, both the algorithms and the interpretations can be expected to be flawed at least on occasion. In fact, in some cases the interpretations could eventually be shown to be wrong not just in minor details but in the full picture presented. For instance, if other chemicals in the atmosphere are causing significant amounts of the radiation adjustments being attributed to ozone, the calculations for the amount of atmospheric ozone might be seriously in error. Similarly, if widespread surface features on sea ice floes produce radiative signals close to those for ice-free water, the calculations of ice concentration could be seriously in error. In both of these cases, as in others, large errors are presumed not to be occurring; but the users of satellite data should always be aware of such possibilities, and aware that future improvements to the algorithms could result in revisions to the earlier data sets and to the details of some of the conclusions drawn from them.

(b) In visible and near-infrared images, clouds can badly obscure the surface data (e.g., Figure 1.5), so that under overcast conditions high-quality data are generally not available for surface variables. This limitation makes the collection of usable surface data highly dependent on weather conditions at the time of the observation. It also prevents the collection (from visible and near-infrared measurements) of a uniform long-term data set on which to attempt global change analyses for any surface variables.

(c) In visible images, an additional limitation is the requirement for visible light, most of which in the Earth's vicinity comes from solar radiation. In the polar regions, this requirement means that usable visible data generally cannot be collected during the polar night, which lasts for months at a time in the near-polar vicinity. Elsewhere also, visible data of standard climate variables are valuable only for periods of sunlight, but outside the polar regions such periods occur every day. Hence, in middle and low latitudes, the need for sunlight does not prevent the collection of a data record that includes data from every day of the year. It does, however, prevent the observation of many low-latitude and mid-latitude nighttime events and the determination of day-night cycles in low-latitude and mid-latitude variables. Still, despite their failure to provide information on many climate variables, nighttime visible images can nonetheless be spectacular, revealing lightning, auroras, or the distribution of cities and other regions of human habitation through the glare of city lights. Major fires, whether natural or manmade, can also be seen on nighttime visible images.

(d) As is the case for visible light, almost all ultraviolet light in the Earth's vicinity is from solar radiation (as discussed in Chapter 2), so that measurements at ultraviolet wavelengths are basically restricted to sunlit areas. This restriction hinders, for instance, the monitoring of stratospheric ozone levels (and the study of the possibility of ozone depletion and ozone holes) in the north polar region in the midst of the Northern Hemisphere winter and in the south polar region in the midst of the Southern Hemisphere winter (Chapter 3), and it accounts for the pole-centered areas of missing data in the TOMS ozone images (Figures 3.5–3.7).

(e) The relatively coarse spatial resolution of passive-microwave images (versus ultraviolet, visible, or infrared images) produces a different set of problems as a result of the greater variety of spatial surfaces and atmospheric conditions contained within the field of view from which a single point of satellite data is collected. Typically, the field of view of a satellite passive-microwave instrument has a diameter between 1 kilometer and 500 kilometers, with only one radiation value being recorded at each wavelength over the entire field of view. Hence, for instance, within the passive-microwave instrument's field of view at any particular moment, if a snow cover exists it almost certainly has varying thicknesses, if sea ice exists it almost certainly has varying concentrations and types, and if vegetation exists it almost certainly includes different species and densities. When an algorithm yields a value for a variable such as snow thickness, the aim generally is to provide a reasonable average value for the full field of view; there is no possibility, with the satellite instrument, of obtaining the spatial distribution of snow thicknesses internal to the field of view.

Furthermore, all along the land–ocean boundary, a satellite instrument's field of view generally contains some land and some ocean. Hence, the ocean variables become contaminated with land radiation and the land variables become contaminated with ocean radiation, giving rise to the terms "land contamination" and "ocean contamination," respectively. This problem is discussed in the chapter on sea ice (Chapter 4), in which the ice concentration images contain noticeable land contamination along much of the continental boundary. Such land contamination exists in all passive-microwave sea ice images except where it has been artificially removed, in which event it is important for the user to know what the removal scheme entailed and how it might affect the rest of the data. Land and ocean contamination both exist immediately along the coast for high-resolution data as well, but in those cases the contamination is less serious because it extends a much shorter distance from the coastline.

(f) Irrespective of which wavelengths are employed in making the measurements, some surface variables interfere with others. For instance, current methods of measuring vegetation fail when a snow cover overlies the vegetation, and, conversely, current satellite-based measurements of snow thickness are rendered inaccurate when vegetation, buildings, and other constructs stick through the snow cover over a sizable portion of the field of view.

(g) Also irrespective of which wavelengths are employed, any measurements of surface variables are complicated by the presence of the atmosphere between the surface and the satellite. The complications are decidedly greater for some wavelengths and atmospheric conditions than for others—for instance, clouds produce major problems at visible and near-infrared wavelengths, as noted earlier in this section—but they occur to some extent for all satellite measurements of the Earth's surface.

Another very important caution regarding satellite data is the full suite of difficulties that arise during the analysis of time series and the determination of geophysical trends, such as warming or cooling trends or trends toward increased or decreased coverage of sea ice, glacier ice, or continental snow. Many instruments physically change over time, resulting in a "drift" in the recorded radiation values that is caused by the instrument changes but can easily be misinterpreted as resulting from a trend in the variable being observed. In any individual case, determining whether an observed drift is an instrument drift or a geophysical trend or other geophysical change can be extremely difficult, sometimes generating major disagreements amongst informed scientists working together on the problem. The situation becomes even more difficult when a data record derives from more than one instrument. When new instruments are launched to extend the record of a current instrument, the algorithms often require fine-tuning to get the records of the two instruments to match reasonably well. Even when the instruments are identical, results from them can differ for a variety of reasons, including differing altitudes or orbits of the satellites and differing times of day of the observations. Sorting out the differences can be time-consuming and difficult, whereas failing to do so can lead to serious misinterpretations of the satellite record. Likely there have been at least a few instances in which changes reported as being changes in geophysical variables were actually caused by changes in instruments; and likely additional such instances will occur in the future as well.

Review Questions

1. Although satellite measurements can provide information on many different variables, the satellite instruments actually directly measure only one thing (with many types or categories). What is that one thing?

2. What is the general term used for a set of equations that converts from the values measured directly by a satellite instrument to such quantities as sea ice concentrations, atmospheric ozone amounts, or sea surface temperatures?

3. Should the user of geophysical satellite data products assume that the satellite images are perfect?

4. Are visible data more reliable for obtaining long-term, consistent data sets of cloud cover or of vegetation?

5. **a.** Are nighttime visible images in June most likely to be of value in examining natural features in the Arctic, the Antarctic, or the equatorial regions?
   b. Why?

6. Which of the following comes closest to explaining why TOMS ozone images routinely show missing data over Antarctica in June and July? (a) The Antarctic ozone hole prevents the collection of ozone data. (b) There is insufficient sunlight to allow the measurements to be made. (c) The orbit of the satellite prevents the collection of data over Antarctica. [*Note:* Two of these statements are false; the other is the correct answer.]

7. Why are land contamination and ocean contamination more serious problems when using data from an instrument with a coarse spatial resolution (having a large field of view) than when using data from an instrument with a much finer spatial resolution (having a smaller field of view)?

8. How can instrument drift confuse the search for geophysical trends?

9. When an identical copy of an existing satellite instrument is launched into space to continue the record begun by the first instrument, should the two instruments be expected to obtain identical results during the period when both are operational?

Looking Toward the Future

In view of the value of satellite observations and advances in technology, it is not surprising that the number of observations, the number of variables being examined, and the number of countries participating have all increased tremendously since the launch of the first satellites by the Soviet Union and the United States in the late 1950s. Also not surprisingly, plans are underway that should lead to further increases in each of these numbers in the years ahead. At the same time, work is progressing toward improvements in the quality of the observations and in the quality of the products derived from them, addressing many of the current limitations discussed in the previous section. Having been in existence for only a few decades, satellite technology is still a fairly new field, and advances continue to come at a fairly rapid pace. Further improvements are expected in several arenas: better instruments, better algorithms to convert the measurements to desired geophysical information, longer and more consistent records, and records of additional variables not yet being measured from space.

Two sample instrumental improvements planned for the early part of the twenty-first century are (1) the use of laser technology in Earth-observing satellite altimeters, markedly improving the surface-elevation accuracies obtainable over those from the radar altimeters of the 1970s through the 1990s, and (2) vast increases in the number of channels on satellite infrared sounders, yielding air temperatures at more levels within the atmosphere and at greater accuracies.

Regarding algorithms, two sample categories of planned improvements are (1) the incorporation of more sophisticated schemes for correcting for atmospheric conditions when deriving information on surface variables, and (2) development of algorithms that are more physically based, to replace some of the algorithms that remain basically empirical.

Some advances will almost assuredly come serendipitously. A key example of such advances from the past is the development of the use of data from the Total Ozone Mapping Spectrometer (TOMS) to examine volcanic emissions. As discussed in Chapter 8, the first TOMS satellite instrument was launched in 1978, and the data from it were being successfully used to provide information on atmospheric ozone when, in early spring 1982, the ozone calculations showed unrealistically high ozone amounts over Mexico. It quickly became apparent that these were false indications of ozone and that the TOMS instrument was somehow tracking emissions from the March 28–April 4 El Chichón volcanic eruption. Further examination led to a determination of the cause: sulfur dioxide in the emissions affected the ultraviolet radiation reaching the TOMS in a way that was being interpreted in the calculations as having been due to ozone. The false ozone values revealed an imperfection in the ozone calculations but at the same time revealed a wonderful new application of the TOMS data. Suddenly it became possible to monitor from space, through the TOMS instrument, emissions from volcanic eruptions. Considerable work followed, including the development of algorithms for sulfur dioxide determinations, the examination of past TOMS data for records of earlier eruptions, and the setting up of procedures to monitor the emissions from future eruptions as they occur. Hence an algorithm imperfection for one variable, ozone, once understood, was converted into a giant step in the satellite-derived measurement of another variable, namely, volcanic sulfur dioxide emissions. Since the El Chichón eruption, the TOMS has continued to be used for volcano studies as well as ozone studies, and, in fact, in some instances eruptions in remote regions have been detected through the TOMS observations before they have become known through any other means.

The TOMS example of an important role played by serendipity in satellite Earth observations is by no means unique. Another example mentioned in this book (in Chapter 7) concerns applications of data from the Advanced Very High Resolution Radiometer (AVHRR). Specifically, the AVHRR instrument was modified in 1979 to improve the ability to distinguish clouds from land, and later those modifications were recognized as opening new opportunities to determine measures of land vegetation. These opportunities have subsequently led to major new land-vegetation data sets.

The TOMS, AVHRR, and similar examples are instructive for several reasons. Amongst the most important, they highlight the facts that algorithms for geophysical products are imperfect, that being open to those imperfections can lead to major advances in unexpected directions, and that to accomplish the advances requires considerable additional work, in the TOMS and AVHRR cases through the development and implementation of algorithms for measures of sulfur dioxide and land vegetation, respectively.

No algorithm or mathematical formulation of any kind comes close to incorporating the full complexity of the Earth system. The algorithms used to obtain the satellite images presented in this book range considerably in level of sophistication, with the ozone calculations being amongst the most complicated and the snow calculations amongst the simplest. In the case of the snow calculations, the entire algorithm consists simply of subtracting two radiative values and multiplying the result by a constant. More complicated snow algorithms exist, but even the simple one used in Chapter 5 yields considerable information about large-scale snow coverage through the year and from year to year. Hence, a person need not wait for a perfect algorithm (which will likely never exist) before using the satellite data to derive information of value. He or she should, however, always be open to the possibility of improvements. Technological, theoretical, and empirical developments all have the potential of leading to improved algorithms and thereby to improved data sets.

To conclude, the field of satellite observations has produced magnificent new views of the Earth, yet it remains full of additional potential, with ample opportunities for significant new progress not just by today's researchers but by today's students and, eventually, the students of those students. Opportunities exist to improve instruments, to improve algorithms, and of course to improve the analyses of the lengthening data sets and thereby to obtain a more complete understanding of the Earth–atmosphere system.

Review Questions

1. Which of the following have increased since the late 1950s: the number of satellite observations of the Earth, the number of variables being examined through satellite observations of the Earth, the number of countries participating in satellite observations of the Earth?

2. **a.** Of satellite laser altimeters and satellite radar altimeters, which were used for Earth observations in the 1980s?
   b. Which are expected to yield the greater surface-elevation accuracies?

3. Explain how the 1982 El Chichón volcanic eruption led to a major new application of TOMS data.

4. Color-coded satellite images have been provided in this book for the following variables—atmospheric ozone, sea ice, continental snow, sea surface temperature, land vegetation, and volcanic emissions. For which of these variables are improved algorithms likely to be developed within the next two decades?

Acronyms and Abbreviations
(including chemical symbols)

AVHRR	Advanced Very High Resolution Radiometer
°C	Degrees centigrade
CCl_4	Carbon tetrachloride
CD-ROM	Compact Disk-Read Only Memory
CFC	Chlorofluorocarbon
CH_3CCl_3	Methyl chloroform
$C_6H_{12}O_6$	Carbohydrate
CH_3SCH_3	Dimethylsulfide
CLAES	Cryogenic Limb Array Etalon Spectrometer
CO_2	Carbon dioxide
DAAC	Distributed Active Archive Center
DMS	Dimethylsulfide
DMSP	Defense Meteorological Satellite Program
ENSO	El Niño/Southern Oscillation
ERBS	Earth Radiation Budget Satellite
ESMR	Electrically Scanning Microwave Radiometer
GAC	Global Area Coverage
GCM	General Circulation Model
GIMMS	Global Inventory Monitoring and Modeling Studies
GMT	Greenwich Mean Time
HALOE	Halogen Occultation Experiment
H_2O	Water
H_2SO_4	Sulfuric acid
ISLSCP	International Satellite Land Surface Climatology Project
JPL	Jet Propulsion Laboratory
K	Kelvin (a unit of temperature)
MCSST	Multichannel Sea Surface Temperature
MLS	Microwave Limb Sounder

MSS	Multispectral Scanner
NASA	National Aeronautics and Space Administration
NDVI	Normalized Difference Vegetation Index
NOAA	National Oceanic and Atmospheric Administration
NSIDC	National Snow and Ice Data Center
O	Oxygen atom
O_2	Oxygen molecule
O_3	Ozone
OH	Hydroxyl radical
PODAAC	Physical Oceanography Distributed Active Archive Center
SAGE	Stratospheric Aerosol and Gas Experiment
SAR	Synthetic Aperture Radar
SI	Système International
SMMR	Scanning Multichannel Microwave Radiometer
SO_2	Sulfur dioxide
SSMI	Special Sensor Microwave Imager
SST	Sea Surface Temperature
TIROS	Television and Infrared Observation Satellite
TM	Thematic Mapper
TOMS	Total Ozone Mapping Spectrometer
UARS	Upper Atmosphere Research Satellite
URL	Uniform Resource Locator
UV	Ultraviolet
WWW	World Wide Web

Additional Reading

Chapter 1. Introduction: Visible Images from Space

Baker, D. J., 1990: *Planet Earth: The View from Space*, Harvard University Press, Cambridge, Massachusetts, 191 pp.

Burroughs, W. J., 1991: *Watching the World's Weather*, Cambridge University Press, Cambridge, England, 196 pp.

Stevens, P. R. and K. W. Kelley, 1992: *Embracing Earth: New Views of Our Changing Planet*, Chronicle Books, San Francisco, 176 pp.

Chapter 2. Radiation

Barry, R. G. and R. J. Chorley, 1985: Solar radiation, Surface receipt of solar radiation and its effects, and Infrared radiation from the Earth, in *Atmosphere, Weather and Climate*, 4th edition, Methuen, London, pp. 9–33.

Kiehl, J. T., 1996: Black body and grey body radiation, in *Encyclopedia of Climate and Weather*, edited by S. H. Schneider, Oxford University Press, New York, vol. 1, pp. 92–93.

Platt, C. M. R., 1996: Radiation, in *Encyclopedia of Climate and Weather*, edited by S. H. Schneider, Oxford University Press, New York, vol. 2, pp. 629–634.

Rees, W. G., 1990: *Physical Principles of Remote Sensing*, Cambridge University Press, Cambridge, England, 247 pp.

Sobel, M. I., 1987: *Light*, University of Chicago Press, Chicago, 263 pp.

Tobias, C. A. and J. Silverman, 1989: Radiation, in *The New Encyclopaedia Britannica: Macropaedia, Knowledge in Depth*, 15th edition, Editor in Chief P. W. Goetz, Encyclopaedia Britannica, Chicago, vol. 26, pp. 492–521.

Waldman, G., 1983: *Introduction to Light: The Physics of Light, Vision, and Color*, Prentice-Hall, Englewood Cliffs, New Jersey, 228 pp.

Chapter 3. Atmospheric Ozone and the Antarctic Ozone Hole

Albritton, D. L. and R. Monastersky, 1992: *Our Ozone Shield*, no. 2 in the series *Reports to the Nation on Our Changing Planet*, University Corporation for Atmospheric Research, 21 pp.

De Gruijl, F. R., 1995: Impacts of a projected depletion of the ozone layer, *Consequences*, vol. 1, no. 2, pp. 12–21.

Ehrlich, P. R. and A. H. Ehrlich, 1991: Ozone: A cautionary tale, in *Healing the Planet: Strategies for Resolving the Environmental Crisis*, Addison-Wesley, Reading, Massachusetts, pp. 113–129.

Ennis, C. A., coordinating editor, 1995: *Scientific Assessment of Ozone Depletion: 1994: Executive Summary*, World Meteorological Organization Global Ozone

Research and Monitoring Project, Report No. 37, 36 pp. (Available from the United Nations Environment Programme, Ozone Secretariat, P. O. Box 30552, Nairobi, Kenya, or from the WMO Global Ozone Observing System, P. O. Box 2300, 1211-Geneva-2, Switzerland.)

Farman, J. C., B. G. Gardiner, and J. D. Shanklin, 1985: Large losses of total ozone in Antarctica reveal seasonal ClO_x/NO_x interaction, *Nature*, vol. 315, no. 6016, pp. 207–210.

Lloyd, S. A., 1993: Stratospheric ozone depletion, *Lancet*, vol. 342, pp. 1156–1158.

Molina, M. J. and F. S. Rowland, 1974: Stratospheric sink for chlorofluoromethanes: Chlorine atom-catalysed destruction of ozone, *Nature*, vol. 249, no. 5460, pp. 810–812.

Pyle, J. A., G. Carver, J. L. Grenfell, J. A. Kettleborough, and D. J. Lary, 1992: Ozone loss in Antarctica: The implications for global change, *Philosophical Transactions of the Royal Society of London*, Series B, vol. 338, no. 1285, pp. 219–226.

Rowland, F. S., 1989: Chlorofluorocarbons, stratospheric ozone, and the Antarctic 'ozone hole,' in *Global Climate Change: Human and Natural Influences*, edited by S. F. Singer, Paragon, New York, pp. 113–155.

Rowlands, I. H., 1993: The fourth meeting of the parties to the Montreal Protocol: Report and reflection, *Environment*, vol. 35, no. 6, pp. 25–34.

Schoeberl, M. R., 1993: Stratospheric ozone depletion, in *Atlas of Satellite Observations Related to Global Change*, edited by R. J. Gurney, J. L. Foster, and C. L. Parkinson, Cambridge University Press, Cambridge, England, pp. 59–65.

Solomon, S., 1990: Progress towards a quantitative understanding of Antarctic ozone depletion, *Nature*, vol. 347, pp. 347–354.

Stolarski, R. S., 1988: The Antarctic ozone hole, *Scientific American*, vol. 258, no. 1, pp. 30–36.

Chapter 4. Polar Sea Ice

Barry, R. G., M. C. Serreze, J. A. Maslanik, and R. H. Preller, 1993: The Arctic sea ice-climate system: Observations and modeling, *Reviews of Geophysics*, vol. 31, no. 4, pp. 397–422.

Brigham, L. W., editor, 1991: *The Soviet Maritime Arctic*, Naval Institute Press, Annapolis, Maryland, 336 pp.

Fogg, G. E., 1992: *A History of Antarctic Science*, Cambridge University Press, Cambridge, England, 483 pp.

Gloersen, P., W. J. Campbell, D. J. Cavalieri, J. C. Comiso, C. L. Parkinson, and H. J. Zwally, 1992: *Arctic and Antarctic Sea Ice, 1978–1987: Satellite Passive-Microwave Observations and Analysis*, NASA SP-511, National Aeronautics and Space Administration, Washington, D.C., 290 pp.

Hall, D. K. and J. Martinec, 1985: *Remote Sensing of Ice and Snow*, Chapman and Hall, London, 189 pp.

Massom, R., 1991: *Satellite Remote Sensing of Polar Regions*, Lewis Publishers, Boca Raton, Florida, 307 pp.

Parkinson, C. L., 1989: On the value of long-term satellite passive-microwave data sets for sea ice/climate studies, *GeoJournal*, vol. 18, no. 1, pp. 9–20.

Parkinson, C. L., 1996: Sea ice, in *Encyclopedia of Climate and Weather*, edited by S. H. Schneider, Oxford University Press, New York, vol. 2, pp. 669–675.

Parkinson, C. L., J. C. Comiso, H. J. Zwally, D. J. Cavalieri, P. Gloersen, and W. J. Campbell, 1987: *Arctic Sea Ice, 1973–1976: Satellite Passive-Microwave Observations*, NASA SP-489, National Aeronautics and Space Administration, Washington, D.C., 296 pp.

Zwally, H. J., J. C. Comiso, C. L. Parkinson, W. J. Campbell, F. D. Carsey, and P. Gloersen, 1983: *Antarctic Sea Ice, 1973–1976: Satellite Passive-Microwave Observations*, NASA SP-459, National Aeronautics and Space Administration, Washington, D.C., 206 pp.

Chapter 5. Continental Snow Cover

Chang, A. T. C., J. L. Foster, and D. K. Hall, 1987: Nimbus-7 SMMR derived global snow cover parameters, *Annals of Glaciology*, vol. 9, pp. 39–44.

Clarke, G. K. C., 1987: A short history of scientific investigations on glaciers, *Journal of Glaciology*, special issue, pp. 1–21.

Foster, J. L., 1983: Night-time observations of snow using visible imagery, *International Journal of Remote Sensing*, vol. 4, no. 4, pp. 785–791.

Foster, J. L. and A. T. C. Chang, 1993: Snow cover, in *Atlas of Satellite Observations Related to Global Change*, edited by R. J. Gurney, J. L. Foster, and C. L. Parkinson, Cambridge University Press, Cambridge, England, pp. 361–370.

Hall, D. K. and J. Martinec, 1985: *Remote Sensing of Ice and Snow*, Chapman and Hall, London, 189 pp.

Kunzi, K. F., S. Patil, and H. Rott, 1982: Snow-cover parameters retrieved from Nimbus-7 Scanning Multichannel Microwave Radiometer (SMMR) data, *IEEE Transactions on Geoscience and Remote Sensing*, vol. GE-20, no. 4, pp. 452–467.

Male, D. H., 1980: The seasonal snowcover, in *Dynamics of Snow and Ice Masses*, edited by S. C. Colbeck, Academic Press, New York, pp. 305–395.

Matson, M., C. F. Ropelewski, and M. S. Varnadore, 1986: *An Atlas of Satellite-Derived Northern Hemisphere Snow Cover Frequency*, National Weather Service, Washington, D.C., 75 pp.

Paterson, W. S. B., 1994: *The Physics of Glaciers*, 3rd edition, Pergamon, New York, 480 pp.

Robinson, D. A., K. F. Dewey, and R. R. Heim, Jr., 1993: Global snow cover monitoring: An update, *Bulletin of the American Meteorological Society*, vol. 74, no. 9, pp. 1689–1696.

Walsh, J. E., 1984: Snow cover and atmospheric variability, *American Scientist*, vol. 72, no. 1, pp. 50–57.

Chapter 6. Sea Surface Temperatures and the El Niño

Andreae, M. O., 1987: The oceans as a source of biogenic gases, *Oceanus*, vol. 29, no. 4, pp. 27–35.

Gedzelman, S. D., 1995: Our global perspective, *Weatherwise*, vol. 48, no. 3, pp. 63–67.

Glantz, M.H., 1996: *Currents of Change: El Niño's Impact on Climate and Society*, Cambridge University Press, Cambridge, England, 194 pp.

Hogg, N., 1992: The Gulf Stream and its recirculations, *Oceanus*, vol. 35, no. 2, pp. 18–24.

Kidwell, K., 1995: *NOAA Polar Orbiter Data Users Guide*, U. S. Department of Commerce, NOAA/National Environmental Satellite Data and Information Service, National Climatic Data Center, Washington, D.C., 410 pp.

Lau, K.-M. and A. J. Busalacchi, 1993: El Niño Southern Oscillation: A view from space, in *Atlas of Satellite Observations Related to Global Change*, edited by R. J. Gurney, J. L. Foster, and C. L. Parkinson, Cambridge University Press, Cambridge, England, pp. 281–294.

Leetmaa, A., 1989: The interplay of El Niño and La Niña, *Oceanus*, vol. 32, no. 2, pp. 30–34.

Njoku, E. G. and O. B. Brown, 1993: Sea surface temperature, in *Atlas of Satellite Observations Related to Global Change*, edited by R. J. Gurney, J. L. Foster, and C. L. Parkinson, Cambridge University Press, Cambridge, England, pp. 237–249.

Philander, S. G., 1990: *El Niño, La Niña, and the Southern Oscillation*, Academic Press, San Diego, California, 293 pp.

Philander, S. G., 1992: El Niño, *Oceanus*, vol. 35, no. 2, pp. 56–61.

Takahashi, T., 1989: The carbon dioxide puzzle, *Oceanus*, vol. 32, no. 2, pp. 22–29.

Vazquez, J., A. van Tran, R. Sumagaysay, E. Smith, and M. Hamilton, 1995: *NOAA/NASA AVHRR Oceans Pathfinder Sea Surface Temperature Data Set User's Guide*, version 1.2, September 27, 1995, Jet Propulsion Laboratory, California Institute of Technology, Pasadena, California. (Available over the world wide web at http://podaac-www.jpl.nasa.gov/sst.)

Wallace, J. M. and S. Vogel, 1994: *El Niño and Climate Prediction*, no. 3 in the series *Reports to the Nation on Our Changing Planet*, University Corporation for Atmospheric Research, Boulder, Colorado, 25 pp.

Chapter 7. Land Vegetation

Arritt, S., 1993: *The Living Earth Book of Deserts*, Reader's Digest Association, Pleasantville, New York, 224 pp.

Choudhury, B. J., 1993: Desertification, in *Atlas of Satellite Observations Related to Global Change*, edited by R. J. Gurney, J. L. Foster, and C. L. Parkinson, Cambridge University Press, Cambridge, England, pp. 313–325.

Ehrlich, P. R. and A. H. Ehrlich, 1991: *Healing the Planet: Strategies for Resolving the Environmental Crisis*, Addison-Wesley, Reading, Massachusetts, 366 pp.

Gabler, R. E., R. J. Sager, and D. L. Wise, 1994: Ecosystems, *Essentials of Physical Geography*, 4th edition, Harcourt Brace College Publishers, Fort Worth, Texas, pp. 272–307.

Gates, D. M., 1993: *Climate Change and Its Biological Consequences*, Sinauer Associates, Sunderland, Massachusetts, 280 pp.

Goward, S. N., 1989: Satellite bioclimatology, *Journal of Climate*, vol. 2, no. 7, pp. 710–720.

Gutman, G., D. Tarpley, A. Ignatov, and S. Olson, 1995: The enhanced NOAA global land dataset from the Advanced Very High Resolution Radiometer, *Bulletin of the American Meteorological Society*, vol. 76, no. 7, pp. 1141–1156.

Lanzara, P. and M. Pizzetti, 1995: *Simon & Schuster's Guide to Trees*, Simon & Schuster, New York, 384 pp.

Los, S. O., C. O. Justice, and C. J. Tucker, 1994: A global 1° by 1° NDVI data set for climate studies derived from the GIMMS continental NDVI data, *International Journal of Remote Sensing*, vol. 15, no. 17, pp. 3493–3518.

Sellers, P. J., B. W. Meeson, F. G. Hall, G. Asrar, R. E. Murphy, R. A. Schiffer, F. P. Bretherton, R. E. Dickinson, R. G. Ellingson, C. B. Field, K. F. Huemmrich, C. O. Justice, J. M. Melack, N. T. Roulet, D. S. Schimel, and P. D. Try, 1995: Remote sensing of the land surface for studies of global change: Models—algorithms—experiments, *Remote Sensing of Environment*, vol. 51, no. 1, pp. 3–26.

Skole, D. and C. Tucker, 1993: Tropical deforestation and habitat fragmentation in the Amazon: Satellite data from 1978 to 1988, *Science*, vol. 260, no. 5116, pp. 1905–1910.

Strahler, A. H. and A. N. Strahler, 1992: World patterns of natural vegetation, in *Modern Physical Geography*, 4th edition, John Wiley & Sons, New York, pp. 516–540.

Townshend, J. R. G., C. J. Tucker, and S. N. Goward, 1993: Global vegetation mapping, in *Atlas of Satellite Observations Related to Global Change*, edited by R. J. Gurney, J. L. Foster, and C. L. Parkinson, Cambridge University Press, Cambridge, England, pp. 301–311.

Tucker, C. J., W. W. Newcomb, and H. E. Dregne, 1994: AVHRR data sets for determination of desert spatial extent, *International Journal of Remote Sensing*, vol. 15, no. 17, pp. 3547–3565.

Wilson, E. O., 1992: *The Diversity of Life*, Belknap Press, Cambridge, Massachusetts, 424 pp.

Chapter 8. Volcanoes

Bluth, G. J. S., S. D. Doiron, C. C. Schnetzler, A. J. Krueger, and L. S. Walter, 1992: Global tracking of the SO_2 clouds from the June, 1991 Mount Pinatubo eruptions, *Geophysical Research Letters*, vol. 19, no. 2, pp. 151–154.

Casadevall, T. J., 1994: *Volcanic Ash and Aviation Safety*, United States Geological Survey Bulletin 2047, U. S. Government Printing Office, Washington, D.C., 450 pp.

Chester, D., 1993: *Volcanoes and Society*, Edward Arnold Press, Sevenoaks, England, 351 pp.

Doiron, S. D., G. J. S. Bluth, C. C. Schnetzler, A. J. Krueger, and L. S. Walter, 1991: Transport of Cerro Hudson SO_2 clouds, *EOS Transactions*, vol. 72, no. 45, pp. 489, 498.

Francis, P., 1993: *Volcanoes: A Planetary Perspective*, Clarendon Press, Oxford University Press, New York, 443 pp.

Francis, P. and S. Self, 1983: The eruption of Krakatau, *Scientific American*, vol. 249, no. 5, pp. 172–187.

Grove, N., 1992: Volcanoes: Crucibles of creation, *National Geographic*, vol. 182, no. 6, pp. 5–41.

Krueger, A. J., L. S. Walter, P. K. Bhartia, C. C. Schnetzler, N. A. Krotkov, I. Sprod, and G. J. S. Bluth, 1995: Volcanic sulfur dioxide measurements from the total ozone mapping spectrometer instruments, *Journal of Geophysical Research*, vol. 100, no. D7, pp. 14,057–14,076.

McCredie, S., 1994: When nightmare waves appear out of nowhere to smash the land, *Smithsonian*, vol. 24, no. 12, pp. 28–39.

McGuire, B., C. Kilburn, and J. Murray, editors, 1995: *Monitoring Active Volcanoes: Strategies, Procedures, and Techniques*, UCL Press, London, 421 pp.

Mouginis-Mark, P. J., D. C. Pieri, and P. W. Francis, 1993: Volcanoes, *Atlas of Satellite Observations Related to Global Change*, edited by R. J. Gurney, J. L. Foster, and C. L. Parkinson, Cambridge University Press, Cambridge, England, pp. 341–357.

Officer, C. and J. Page, 1993: *Tales of the Earth: Paroxysms and Perturbations of the Blue Planet*, Oxford University Press, New York, 226 pp.

Rampino, M. R. and S. Self, 1984: The atmospheric effects of El Chichón, *Scientific American*, vol. 250, no. 1, pp. 48–57.

Simkin, T. and R. S. Fiske, 1983: *Krakatau 1883: The Volcanic Eruption and Its Effects*, Smithsonian Institution Press, Washington, D.C., 464 pp.

Singer, S. F., editor, 1989: *Global Climate Change: Human and Natural Influences*, Chapters 20 (D. Lal, Global effects of meteorite impacts and volcanism, pp. 339–362), 21 (A. A. Lacis and S. Lebedeff, Commentary on Devendra Lal's paper, pp. 363–371), 22 (H. W. Ellsaesser, Further commentary on Devendra Lal's paper, pp. 373–376), and 23 (H. G. Goodell, Asteroids, volcanoes, and climate, pp. 377–383), Paragon House, New York, 424 pp.

Stommel, H. and E. Stommel, 1983: *Volcano Weather*, Seven Seas Press, Newport, Rhode Island, 177 pp.

Chapter 9. Conclusions: Strengths and Limitations of Satellite Data

Asrar, G. and D. J. Dokken, editors, 1995: *The State of Earth Science from Space: Past Progress, Future Prospects*, American Institute of Physics Press, Woodbury, New York, 159 pp.

Asrar, G. and J. Dozier, 1994: *Science Strategy for the Earth Observing System*, American Institute of Physics Press, Woodbury, New York, 119 pp.

Gurney, R. J., J. L. Foster, and C. L. Parkinson, editors, 1993: *Atlas of Satellite Observations Related to Global Change*, Cambridge University Press, Cambridge, England, 470 pp.

Stevens, P. R. and K. W. Kelley, 1992: *Embracing Earth: New Views of Our Changing Planet*, Chronicle Books, San Francisco, 176 pp.

Answers

Chapter 1. Introduction: Visible Images from Space

1. **a.** Africa's.

   b. Antarctica's coastline is less clearly delineated for two major reasons: (1) The fact that the clouds and the Antarctic continent are both white makes it difficult at some locations to distinguish between the two. (2) At the moment this photograph was taken, a greater percentage of the Antarctic coastline than the African coastline was overlain by clouds.

2. Figure 1.1.
3. Figure 1.4.
4. Visible radiation (or visible light).
5. The visible images are easy to understand.
6. Cloudy conditions; dark conditions.
7. Clear daytime conditions.
8. Cloudy daytime conditions.

Chapter 2. Radiation

Introduction

1. There are no familiar objects that do not give off radiation. ("A blackhole" could also be considered a correct answer, although it would be difficult to argue that a blackhole is a familiar object.)
2. 300,000 kilometers per second.
3. The speed of light (or the speed of radiation); 300,000 kilometers per second.
4. The distance from one peak of the wavelike form in which the radiation is traveling to the next peak.
5. **a.** Violet.
   b. Red.
6. The Sun.
7. Longer.

The Electromagnetic Spectrum

1. In order: X-rays, ultraviolet radiation, violet visible radiation, red visible radiation, infrared radiation, microwave radiation, radio-band radiation.
2. Visible radiation (or, also correct, red radiation).

3. Radiation type B.
4. 9.68 GigaHertz.
5. 30 picometers.

Blackbody Radiation Curves

1. An object that gives off radiation at the maximum rate possible, for its temperature, at each wavelength.
2. No.
3. 373.15 K.
4. **a.** The curve for the object having a temperature of 900 K.
   b. The curve for the object having a temperature of 700 K.
5. The infrared region.
6. Because, as a result of the great distance between the Sun and the Earth, only a fraction of the radiation given off by the Sun actually reaches the Earth, and, in the case of microwave radiation, the amount reaching the Earth from the Sun is much less than what the Earth-atmosphere system gives off.
7. By the Sun.
8. Thermal-infrared wavelengths.

Chapter 3. Atmospheric Ozone and the Antarctic Ozone Hole

Introduction

1. Nitrogen and oxygen.
2. **a.** Oxygen.
   b. Three.
3. No.
4. It provides protection from excessive ultraviolet radiation (by absorbing it).
5. Sunburn, skin cancer, eye damage, increased immune deficiencies (any two).
6. Release of CFCs into the atmosphere through use of aerosol sprays, refrigerants, air conditioners, heat pumps, solvents, and plastic foams. Also, release of halons into the atmosphere through use of fire extinguishers.
7. Through analysis of data from the British ground station at Halley Bay, Antarctica. (The data were from spectrophotometers.)

8. October.

9. Yes (including the Arctic and mid-latitude regions of the Northern Hemisphere).

10. Ice particles.

11. An international agreement to limit further production of ozone-depleting chemicals.

Satellite Detection of Ozone

1. No.

2. The TOMS and several UARS instruments (such as the CLAES and the HALOE).

3. An algorithm is used to convert from radiation values to ozone amounts. The algorithm is based on the fact that ozone absorbs ultraviolet radiation (hence, if high values of ultraviolet radiation are received by the satellite, the ozone amounts in the atmosphere are probably low) and does so to different extents at different wavelengths.

4. $M = 60 \times H$.

5. Dobson units.

6. 0 to 1 centimeter.

7. Other chemical components of the atmosphere that absorb ultraviolet radiation could result in reduced ultraviolet amounts recorded by the TOMS, and these reduced ultraviolet amounts could erroneously be interpreted as arising from elevated levels of ozone. In such a case, these chemical components would result in calculated ozone amounts that would be too high.

8. No; measurements in the polar regions are not useful in the middle of winter because of the lack of solar radiation.

Satellite Images of Ozone

1. The Northern Hemisphere.

2. About 60°N, 50°W–80°W; 465–500 Dobson units.

3. In the central south polar region, 85°S–90°S; 220–255 Dobson units.

4. About 60°S, 95°E eastward to 165°W; 395–430 Dobson units.

5. In the south polar region; 150–185 Dobson units.

6. In order: March, June, September.

7. In order: September, December, March.

8. In both cases, the highest values were in the month at the end of winter and start of spring (March in the Northern Hemisphere and September in the Southern Hemisphere), the middle values were in the month of the summer solstice (June in the Northern Hemisphere, December in the Southern Hemisphere), and the lowest values were in the month at the end of summer, start of autumn (September in the Northern Hemisphere, March in the Southern Hemisphere).

9. a. None.
   b. None.

10. September.

11. October.

12. a. 465–500 Dobson units; about 55°S–65°S and 30°E–130°E.
   b. 430–465 Dobson units.
   c. 360–395 Dobson units.
   d. 465–500 Dobson units.

13. a. 255–290 Dobson units.
   b. 185–220 Dobson units.
   c. 150–185 Dobson units.
   d. 185–220 Dobson units.

14. 1985.

Chapter 4. Polar Sea Ice

Introduction

1. a. Ice formed through the freezing of sea water.
   b. Less.

2. Sea ice is ice formed in the sea from sea water; icebergs are formed by the breaking off into the sea of glacier ice formed on the land from fallen snow. (Sea ice hence is somewhat salty, whereas icebergs are not.)

3. In order: icebergs, sea ice, sea water.

4. Because salt is released from the ice to the water underneath as the ice forms and ages.

5. Salt rejection during ice formation increases the salt content and hence the density of the water directly under the ice. At times, this process causes the water directly under the ice to become denser than the water further down, leading to overturning. When the overturning extends to sufficient depths, deep water is formed.

6. Decrease.

7. It will be increased.

8. The area of North America.

9. Before.

10. Satellite technology.

Satellite Detection of Sea Ice

1. Sea ice.

2. Both instrument types measure microwave radiation, but the passive instrument simply receives (and records or transmits) radiation emitted from elsewhere, whereas the active instrument sends a signal out and receives it back (after reflection).

3. The contrast between the microwave emission from sea ice and the microwave emission from liquid water.

4. a. Water.
   b. Ice.

5. 1.

6. In order: (1) The ice-free oceans; (2) the ice sheets of Greenland and Antarctica; (3) the sea ice covers of the Arctic and Antarctic; (4) the land surfaces of central Europe and the northern United States.
7. 75%.
8. **a.** 45%.
   b. 90%.
   c. 90%.
9. The ice thickness; the ice salinity; whether a snow cover exists; whether there is any melting, and if so, how much. (*Note:* there are other factors as well, but these are the ones listed in the text.)

Satellite Images of Sea Ice
1. **a.** The satellite orbit did not allow data collection from the SMMR instrument poleward of 84.6° latitude.
   b. The TOMS ozone data came from sunlight, so that missing data occur in the regions experiencing a "polar night".
   c. The sea ice data set.
2. **a.** 44%.
   b. 28%.
   c. 76%.
3. **a.** Atmospherically induced noise.
   b. The problem is reduced for atmospherically induced noise because the bothersome atmospheric conditions are transitory and, in general, occur no more than a few days during the month at any individual location. Thus, the monthly average will be affected by them but at a much reduced level (often becoming invisible on the images because of the color being set to light blue in all areas with calculated ice concentrations of 0–12%). In contrast, the problem is not reduced in the case of land contamination because land contamination exists every day at roughly the same strength at each location.
4. **a.** The Arctic Ocean and the Canadian Archipelago.
   b. All of them except the southern Barents Sea.
5. **a.** About 96% (mostly 88–100%).
   b. About 88% (mostly 76–96%).
6. **a.** Hudson Bay.
   b. Hudson Bay is largely surrounded by the North American continent and has much less access to warm waters to the south than either the Sea of Okhotsk or the Bering Sea. Being therefore subject to the continentality effect, it generally has much colder winters than do open-ocean areas at the same latitudes.
7. The eastern portion.
8. The north coast of Alaska.

9. **a.** 75°N.
   b. 80°N.
10. **a.** Northward.
    b. Because the absence of ice in the region suggests that warm water is being transported from the south. (The Norwegian Current is an extension of the Gulf Stream.)
11. The Weddell Sea.
12. **a.** Approximately 10°E.
    b. Approximately 56°S.
13. 88%.
14. The Northern Hemisphere.

Chapter 5. Continental Snow Cover

Introduction
1. Antarctica.
2. Temperature, humidity, and wind.
3. In order: new snow, old snow, firn, glacier ice, water.
4. In order: forests without a snow cover, snow-covered forests, old snow, new snow.
5. It has a cooling effect, decreasing the air temperatures, because snow's unusually high reflectivity results in most of the solar radiation incident on it being reflected back to space and therefore unavailable for heating.
6. It reduces frost penetration.
7. **a.** Avalanches, blizzards, treacherous driving conditions, floods from excessive or rapid melt.
   b. Scenic beauty; locale for recreational activities such as cross-country and downhill skiing; source of melt water for drinking, hydropower, irrigation.
8. The one consisting of old snow.
9. Mid-September; northeastern Russia (Siberia) and northern Alaska.
10. February.
11. In order: summer minimum snow coverage in the Northern Hemisphere; summer minimum snow coverage in the Southern Hemisphere; winter maximum snow coverage in the Southern Hemisphere; winter maximum snow coverage in the Northern Hemisphere.

Satellite Detection of Snow
1. The snow data are obscured whenever a cloud cover lies overhead, and the data sets do not provide estimates of snow thickness and thus cannot be used alone to obtain snow volume estimates. (Other correct answers are also possible.)
2. The visible and infrared record is longer, going back to 1966. Also, it appears to be better for identifying very thin snow coverage.

3. For a moderate snow cover over land, much of the radiation received by the satellite derives from the land under the snow. As the radiation from the land travels upward through the snow, it gets scattered within the snow cover; the thicker the snow, the more radiation is scattered and the less radiation reaches the satellite.

4. The facts that different ground surfaces emit different amounts of radiation and that the amount of radiation emitted by any surface changes as the surface temperature changes.

5. Snow scatters less of the 18 GigaHertz radiation, so that as the snow cover gets thicker (and both the 18 GigaHertz and the 37 GigaHertz radiative values decrease), the decreases for the 37 GigaHertz data are greater than those for the 18 GigaHertz data, so that the difference between the 18 GigaHertz and 37 GigaHertz data gets larger.

6. A basic assumption used in the development of the algorithm—that much of the radiation comes from the underlying ground surface—does not apply over an ice sheet, and an ice sheet exists over most of Greenland.

7. **a.** Less.
   b. The passive-microwave data do not reveal the shallow (<5 centimeters thick) snow cover, whereas the NOAA data do.

8. (c) snow coverage decreased from the first year to the second year.

Satellite Images of Snow Coverage

1. **a.** 28 centimeters.
   b. 16 centimeters.
   c. 52 centimeters.
2. **a.** North-central Russia.
   b. About 40 centimeters (any answer from 30 centimeters to 45 centimeters).
   c. About 64 centimeters (any answer from 58 centimeters to 67 centimeters).
   d. Southwest coast; higher.
3. **a.** Thicker.
   b. Thicker.
   c. Thinner.
4. February and March.
5. **a.** March to April.
   b. April to May.
6. **a.** 1979.
   b. 1979.
   c. 1979.
   d. 1981.
7. **a.** 1981.
   b. 1979.
   c. 1979.
   d. 1981.

8. **a.** February 1979.
   b. February 1979.
   c. February 1981.
9. 1986.
10. 1979.

Chapter 6. Sea Surface Temperatures and the El Niño

Introduction

1. 70%.
2. **a.** Water.
   b. Hundreds of thousands.
   c. Low to high.
   d. Stabilizing.
3. Momentum.
4. Because the Arctic atmosphere in winter is much colder than the water, which is roughly at the freezing point, and heat flow is predominantly from the warmer to the cooler material.
5. Evaporation.
6. Chlorofluorocarbons.
7. The rate at which carbon dioxide is increasing in the atmosphere is considerably less than the difference between the current estimates of the rate at which carbon dioxide is entering the atmosphere and the rate at which it is leaving the atmosphere. At least one of the three estimates (the rate of increase; the rate of carbon dioxide entering the atmosphere; the rate of carbon dioxide leaving the atmosphere) must be wrong.
8. The Gulf Stream. (The North Atlantic Drift is also an appropriate answer, or the combination of the Gulf Stream and the North Atlantic Drift.)
9. **a.** Generally cooler.
   b. Because it would no longer have the warming influence of the water flow from lower latitudes.
10. **a.** Generally cooled.
    b. Generally warmed.
11. **a.** El Niño/Southern Oscillation.
    b. EN (El Niño).
    c. SO (Southern Oscillation).
12. **a.** Non-El Niño episodes.
    b. El Niño episodes.
    c. Non-El Niño episodes.
13. **a.** Because of the abundant vegetation growth following the heavy rains associated with the El Niño.
    b. Because of the damage done to the large Peruvian fishery industry by the El Niño's reduction in the upwelling of nutrients along the Peruvian coast.

Satellite Detection of Sea Surface Temperatures

1. Because the North Atlantic contained particularly heavily traveled shipping routes, and temperature measurements were often made from on board the ships.
2. Visible and infrared data. (The infrared data include both near-infrared and thermal-infrared data.)
3. Near to visible wavelengths.
4. Empirical.
5. High humidity (or clouds) and high aerosol (or volcanic aerosol) concentrations.
6. To create an improved sea surface temperature data set for the period 1981 to the present.
7. **a.** The project can take advantage of the different expertises available at the different institutions. (This is one correct answer; others are also possible.)
   b. Decision-making could be hindered by having too many people involved and by the difficulties of getting all the critical people together to discuss the issues. (As with 7.a, other correct answers are also possible.)
8. A map projection in which the angles between lines on the map are equal to the corresponding angles on the spherical Earth.
9. Partly empirical.

Satellite Images of Sea Surface Temperatures

1. **a.** March 1990; about 16°C.
   b. September 1990; about 24°C.
   c. About 8°C.
   d. For almost everyone living in either midlatitudes or high latitudes, the answer should be that the air temperature differences are greater.
2. **a.** March 1990.
   b. September 1990.
3. **a.** September 1990; about 14°C.
   b. March 1990; about 21°C.
   c. About 7°C.
4. The anomalously warm waters in the northeastern North Atlantic (having been transported by the Gulf Stream from lower latitudes). (These waters are anomalously warm in comparison with sea surface temperatures at comparable latitudes elsewhere, outside the immediate influence of the Gulf Stream.)
5. All four months shown (March, June, September, December 1990).
6. The flow is basically south-to-north, as can be determined from the sea surface temperature maps of Figures 6.4 and 6.5 because cold water typical of the higher southern latitudes extends northward along the southwest African coast.
7. South-to-north.

8. **a.** Along the equator in the eastern Pacific, along the equator in the Indian Ocean, and in the Gulf of Mexico.
   b. In each of the three locations, May 1987 had higher sea surface temperatures than May 1988.
9. **a.** In the equatorial eastern Pacific.
   b. Lower.
   c. Higher.

Chapter 7. Land Vegetation

Introduction

1. For eating, fuel, clothing materials, writing materials, and building materials.
2. Savannas, grasslands, deserts, and tundra.
3. It helps reduce moisture losses.
4. The narrow needles lose much less moisture to the atmosphere than broad leaves under the same conditions, allowing the tree to retain more water (for photosynthesis and to avoid drying out).
5. In order: rainforests, prairies, steppes, deserts.
6. Humans developed agriculture, intentionally cultivating individual types of vegetation.
7. Grassland.
8. Obtaining timber; clearing land.
9. **a.** 56.3 years.
   b. The decade of the 2040s.
10. Yes.
11. They have provided the core substances of many medicines.
12. To recreate a forest in the same location.
13. It reduces the carbon dioxide content of the atmosphere (because atmospheric carbon dioxide is one of the inputs to the photosynthesis process).
14. Yes.
15. **a.** Increase the dilemma.
    b. Reduce the dilemma.
    c. Reduce the dilemma.
    d. Increase the dilemma.
16. **a.** Increase it.
    b. Because it removes some of the roots that help hold the soil in place.
17. Decrease it.
18. **a.** The Sahara Desert.
    b. Depletion of the groundwater supply, overgrazing, excess cultivation, and cutting of firewood along the edges of the desert.

Satellite Detection of Land Vegetation

1. **a.** The Landsat Thematic Mapper.
   b. The NOAA AVHRR and DMSP SSMI.
2. To improve its ability to distinguish clouds from land.
3. Normalized Difference Vegetation Index.

4. a. Red radiation (and visible radiation in general) helps power photosynthesis.
b. Reflecting near-infrared radiation helps prevent overheating.
5. a. Decrease it; because any radiation that gets absorbed doesn't get reflected back in the direction of the satellite instrument and thus doesn't get recorded by the instrument.
b. Increase it.
c. High values.
6. Instances in which the measurements were taken under conditions of a heavy cloud cover, a heavy snow cover, a major flood, or a serious drought.
7. It is used to screen out cloud-obscured data, by eliminating data from locations and times with the thermal-infrared brightness temperatures below a preselected cutoff (273 K for most locations, 285 K for Africa).
8. Because the maximum values are more likely to give a good indication of the vegetation cover in view of the fact that the complicating factors leading to erroneous NDVI values almost all have the effect of lowering the NDVI.
9. Improved algorithms; finer spatial resolution ($0.5° \times 0.5°$ rather than $1° \times 1°$); a longer data set (10 years, 1986–1995, rather than just 1987 and 1988).

Satellite Images of Land Vegetation
1. South America.
2. a. July 1987.
b. July 1987.
c. April 1987.
d. October 1987.
3. a. Summer.
b. Summer.
c. Summer.
4. a. Southern Hemisphere.
b. Southern Hemisphere.
c. Northern Hemisphere.
5. Northward.
6. In January, missing data are indicated over most of Russia because of the wintertime snow cover and the cold wintertime surface temperatures. The cold and snow have retreated northward by April, when there are low-to-moderate values of NDVI over much of southern Russia. By July, the cold and snow have retreated from most of the country almost entirely, and the spring and summertime blossoming of vegetation has created high NDVI values throughout most of the country. Autumn cooling results in lower NDVI values in October, as the vegetation ripens and leaves fall, and also as the surface gets cold and some snow falls in the north.

7. a. Decreased.
b. Decreased.
c. Decreased.
d. Increased.
e. Decreased.
8. 1987.
9. 1988.
10. To the east of the sea.
11. a. All seasons.
b. All seasons.
c. All seasons.

Chapter 8. Volcanoes

Introduction
1. The distance varies with the eruption but sometimes is several thousands of kilometers.
2. Several meters.
3. A water wave created by a volcanic eruption or by an earthquake.
4. Intake of volcanic ash from an eruption plume can cause jet engines to fail.
5. Extremely important; the Earth's atmosphere was probably formed from volcanic emissions and continues to be modified by them.
6. Phosphorus, potassium, calcium, magnesium, and sulfur.
7. Any three of the following: formation of the Earth's atmosphere; new islands; energy from hot magmas; good building materials from pumice flows; obsidian for tools and jewelry; spectacular sunsets; stunning physical settings (e.g., several national parks); preservation of archeological information; or others not listed in the text, such as the knowledge revealed by volcanoes about the interior of the Earth.
8. a. Within several weeks.
b. Several years.
9. a. Cooling, because the materials injected into the upper atmosphere, and the particles they are converted into (especially sulfuric acid), absorb some solar radiation and reflect some back to space, so that less solar radiation reaches the lower atmosphere and the surface.
b. Warming, because of the increased absorption of radiation in the stratosphere.
10. a. Approximately one to two years after the eruption (i.e., in the second year after the eruption).
b. Because within the first few days, the volcanic materials remain relatively localized compared with their extent over the succeeding months.
11. a. A vertically explosive eruption.
b. A sulfur-rich eruption.
c. A low-latitude eruption.

12. Because the El Chichón eruption sent more material, with high sulfur content, into the stratosphere.
13. Benjamin Franklin.
14. Mount Tambora erupted in 1815, and many believe that this eruption was the cause of the cold temperatures the following year (because of the materials injected into the stratosphere, blocking solar radiation from reaching the surface).
15. The eruption of Mount Pinatubo.
16. Dual simulations can be run with the general circulation model, numerically inserting the volcanic debris in one of the simulations and leaving it out in the other. The resulting temperatures can then be subtracted to see how much colder the simulated troposphere is when the volcanic debris is included.

Satellite Detection of Volcanic Emissions
1. a. AVHRR and SAR.
   b. AVHRR, MLS, TOMS, and SAGE.
2. The TOMS.
3. It became clear that the TOMS instrument could be used to measure and track the sulfur dioxide in volcanic plumes, and this tracking now has been done for many eruptions since El Chichón. Also, it led to corrections in the ozone algorithm.
4. a. The ultraviolet region.
   b. Because sulfur dioxide strongly absorbs ultraviolet radiation, so that the instrument will record lower ultraviolet values when the atmosphere contains considerable sulfur dioxide. (The situation is similar with ozone, but sulfur dioxide is a stronger absorber at some wavelengths—for example, 0.3125 and 0.3175 micrometers—and ozone is a stronger absorber at other wavelengths—for example, 0.3312 and 0.3398 micrometers. Such contrasts are valuable in the effort to sort out the amounts of the individual gases.)
5. Any three of the following: ash, aerosols, low Sun angle, heavy cloud cover, significant dispersal of the volcanic plume before the measurements were taken, and conversion of the sulfur dioxide to sulfuric acid before the measurements were taken.
6. First the total volume of sulfur dioxide is calculated by integrating the thicknesses over the area affected; then the mass of sulfur dioxide is calculated by multiplying the calculated volume of sulfur dioxide by the density of sulfur dioxide.
7. The 1991 eruption of Mount Pinatubo.
8. Because the atmosphere has considerable water vapor and carbon dioxide anyway, so that the

amounts added by an eruption do not stand out against the background atmosphere as strongly as the sulfur dioxide amounts do.

Satellite Images of Volcanic Emissions
1. The thickness of a paper clip.
2. About three weeks.
3. The June 19, June 22, and June 25, 1991 images.
4. June 16 image—10°N–20°N; June 19, June 22, and June 25 images—0°N–10°N.
5. a. 220 milli atmospheres centimeter.
   b. 220 milli atmospheres centimeter.
   c. About 150–160 milli atmospheres centimeter.
   d. About 90 milli atmospheres centimeter.
6. The rate of a human driving a car.
7. Yes.
8. About 60 milli atmospheres centimeter for each of the eight dates.
9. a. August 19, 1991.
   b. About 45°S.
10. Over the period August 19–22, 1991.
11. a. Approximately 1 week.
    b. Shorter.
    c. The Mount Pinatubo cloud.

Chapter 9. Conclusions: Strengths and Limitations of Satellite Data

The Scope and Utility of Satellite Earth Observations
1. Any six of the following: solar radiation reaching the Earth's outer atmosphere, longwave radiation emitted by the Earth, atmospheric temperature, land surface temperature, wind speed, aerosols, water vapor, precipitation, clouds, lightning, several chemical constituents in the atmosphere (in addition to ozone), ocean topography, land topography, ocean circulation, ocean biological productivity, evaporation from the ocean, the Greenland ice sheet, the Antarctic ice sheet, mountain glaciers, ice caps. (These are the additional variables listed in the text; others exist as well and could be included in the answer.)
2. Satellites allow monitoring and tracking of the hurricane while it is still far out at sea, often detecting it well before it is reported by more conventional means. Hence, forecasters know of the existence and past history of the hurricane, giving them a base on which to forecast. This improved information base results in hurricanes being far less likely to arrive unexpectedly at a coast. (The distinctive structure of a hurricane allows its easy identification on satellite visible and infrared images. For example, see Figure 1.3.)

3. Humans diverted waters from rivers running into the sea, for the purpose of using the water in irrigating agricultural fields. This diversion resulted in less water entering the sea and consequently shrinkage of the sea.

4. By having environmental problems such as shrinking seas and expanding deforestation visible on satellite images, a greater number and wider range of people can become aware of the problems early on and hence can encourage or participate in efforts to stop, slow, or even reverse the damage.

5. As long as the satellite imagery is available to the people doing the clean-up, it can help by showing where the oil has spread, thereby reducing the time that must be spent on merely locating the oil edge and increasing the chances of finding and removing the majority of the oil.

6. The satellite data allow observations and monitoring of all regions of the Earth's surface and atmosphere, so that theory and explanations no longer have a database restricted to areas and times where humans have physically been and made the observations. Satellite technology allows measurements to be collected globally and to be repeated, sometimes on a quite frequent basis.

Limitations
1. Radiation.
2. An algorithm.
3. No.
4. Cloud cover.
5. **a.** The Arctic.

b. Because the Arctic has sunlight at nighttime in June, whereas the other two regions do not. (Hence, June nighttime visible images for the Arctic will routinely show natural features. In contrast, June nighttime visible images for the Antarctic and the equatorial regions will only occasionally show natural features, such as an aurora in the Antarctic or a forest fire in the equatorial regions.)

6. (b) There is insufficient sunlight to allow the measurements to be made.

7. Because the contaminated region will tend to extend a greater distance from the coast when the resolution is coarse.

8. Instrument drift results in changes over time in the recorded radiation values, which, if not recognized and corrected for, can be misinterpreted as geophysical trends.

9. No.

Looking Toward the Future
1. All three have increased.
2. **a.** Satellite radar altimeters.
   b. Satellite laser altimeters.
3. The El Chichón eruption inserted large quantities of sulfur dioxide into the atmosphere. The sulfur dioxide led to erroneous results in the TOMS ozone calculations, which in turn led to the development of algorithms to obtain sulfur dioxide values from the TOMS data, providing the capability of identifying and monitoring volcanic emissions.
4. All of them.

Index

Note: The answers have purposely not been indexed. Also, geographic place names are generally not indexed to more than one location map.

About the Author

Claire L. Parkinson is a climatologist at NASA's Goddard Space Flight Center, where she uses satellite data and numerical modeling to examine the Earth's climate, with particular emphasis on polar sea ice. She received a bachelor's degree in mathematics from Wellesley College and a doctorate in climatology from Ohio State University. She has done field work in both the Arctic and the Antarctic, has led a project to create an atlas of Arctic sea ice from satellite data, and has coauthored a textbook on climate modeling and coedited a book on satellite observations related to global change. She has also written a history-of-science reference work titled *Breakthroughs: A Chronology of Great Achievements in Science and Mathematics*. As part of her duties at NASA, she is Project Scientist for the Earth Observing System's PM Mission, scheduled for launch in the year 2000.